Environmental Science

Series Editors: R. Allan U. Förstner W. Salomons

Springer
Berlin
Heidelberg
New York
Barcelona
Budapest
Hong Kong
London
Milan
Paris
Santa Clara
Singapore
Tokyo

E. K. Duursma, J. Carroll

Environmental Compartments

Equilibria and Assessment of Processes Between Air, Water, Sediments and Biota

With 164 Figures and 39 Tables

 Springer

Volume Editors

Prof. Dr. Egbert K. Duursma

Res. Les Marguerites
1305 Chemin des Revoires
06320 La Turbie, France

Dr. JoLynn Carroll

Internationale Atomic Energy Agency
Marine Environment Laboratory
B.P. 800
MC 98012 Monaco

ISBN 3-540-61039-1 Springer-Verlag Berlin Heidelberg New York

Library of Congress Cataloging-in-Publication Data

Duursma, E.K.
Environmental compartments: equilibria and assessment of processes between air, water, sediments and biota / E. K. Duursma, J. Carroll. p. cm. Includes bibliographical references and index.
ISBN 3-540-61039-1 (Hardcover). – ISBN 0-387-61039-1 (Hardcover)
1. Environmental chemistry. 2. Chemical equilibrium.
I. Carroll, J. (JoLynn), 1960- . II. Title QD31.2.D88 1996 628.5–dc20 96-23139 CIP

Typesetting: Camera ready by editors/authors
Media conversion: Anne Marie de Grosbois
Cover Design: Atelier Struve & Partner

SPIN: 10497788   32/3136 - 5 4 3 2 1 0 - Printed on acid-free paper

Although having used it frequently in early publications, the word 'important' originates usually from ignorance of the quantitative facts. It contains neither a qualitative nor a quantitative statistical appreciation and might range from significant to negligible (Egbert Duursma).

Preface

In 1957 on board the research vessel Gauss, Prof. Kurt Kalle[1] of the German Hydrographical Institute, posed the fundamental question: How could differences of newly discovered 0.3 mg dissolved organic carbon/l between surface and 3000 metres depth be sustained over the complete water column? Such a difference would involve an amount of 900 g C/m^2, whereas the primary production, the source of this organic matter, only ranged between 30 and 80 g C/m^2/yr. The answer to this question can be found in Appendix I.

Kalle's approach, given during the Geophysical Year 1957-1958, defines the scope of this book: How are phenomena, even of a small scale, correlated to large-scale processes we observe in different compartments of our global environment?

Present-day detailed environmental problems require a solid understanding of the spatial and temporal links which exist between the different environmental compartments. Processes and equilibria, occurring between substances in biota, air, water, land and aquatic sediments are heavily influenced by broad scales of time and space. These time-scales span from seconds to centuries, where space scales range from cubic millimetres to thousands of cubic kilometres.

It is surprising how well global phenomena can be understood when basic findings are extrapolated from small scales. Naturally such extrapolations will suffer from accuracy, but the investigations may nevertheless lead to a better understanding of phenomena and provide the basis for more sophisticated studies.

The presence of DDT in the environment is at present often considered as having only historical value, due to a ban of production in many countries. Nevertheless, DDT and its metabolites are still distributed world-wide at a low but measurable level in air, water, aquatic organisms and man. It is surprising that the level of ΣDDT (sum of DDT and its derivates) in human beings and particularly mother's milk is about equal for people in various continents, and that this level is also similar to that of mussels around Central and South America, when determined on a fat basis. Although the level is below 0.015% of the lethal concentration[2], this indicates that a global distribution still exists. On the causal process some hypotheses are presented, which differ principally

---

1. Prof. Kalle was the famous chemical oceanographer and author of the 'Stoffhaushalt des Meeres' in the book 'Allgemeine Meereskunde' (Dietrich et al. 1975) of which exists also an English edition (Dietrich et al. 1980).

2. The oral LD_{50} of DDT is 200 mg/kg body weight. As later will be demonstrated, ΣDDT levels in human fat (body fat and milk fat) can reach the order of 0.8 mg/kg lipid which equals about 0.03 mg/kg body weight or milk.

from the often dogmatized theory on accumulation of contaminants in food-chains.

The same is more or less true for PCBs which are still in use or stored as waste. These organochlorines are equally widely distributed in the environment, and their concentrations in organisms have reached a similar background level as DDT in several regions. This occurs in spite of the fact that at present only a small fraction of the total world reserves of 1.2 million ton PCBs is spread out in the atmospheric and oceanic compartments.

High-level nuclear wastes continue to accumulate on the planet due to the growing number of nuclear power stations and the reprocessing of nuclear fuel. The stocks predicted for the year 2005 will be $2.1x10^3$ m^3 for the OECD countries in Europe and the Pacific and $1.3x10^3$ m^3 for the OECD countries in North America (IAEA, 1992) with approximately $7x10^{16}$ Bq ($1.9x10^6$ Ci) per m^3. These high-level wastes will remain a potential source of concern unless long-term facilities for storage are secured.

Although these sources of contaminants can be considered as potential environmental time-bombs, science is sufficiently advanced to deal with them, but only, when both scientists and policy-makers take well-reasoned and thoroughly informed actions.

Besides basic information on theories and measurements on the distribution of substances between air, water, sediments and biota, a number of case studies are presented, which concern budgets on radionuclide and metal contaminants in the marine environment and on organochlorines in both the marine environment and atmosphere.

This book is intended to contribute to a better understanding on the equilibria which exist between the substances in the different environmental compartments. It is primarily directed to university education, but also intended to broaden the views of environmental scientists and policy-makers. The content is based on university courses, extended with pertinent case studies of recent environmental problems. To encourage self-study, a number of exercises are incorporated in the text. The answers are presented in Appendix I. For beginning modellers a disk is added, containing a demo model of radionuclide transport from dumped nuclear waste in the Kara Sea, a shallow Arctic sea east of Novaya Zemlya and a demo model, called COSMO-BIO, illustrating the role of biodiversity in coastal zone management, assessing risks of multiple stresses caused at the population level by various human activities.

January 1996   Egbert Klaas Duursma and JoLynn Carroll

Acknowledgments

The authors are very indebted to Dr. Michael L. Carroll who skillfully served as reviewer and English corrector for the book. The help obtained from many colleagues from the IAEA (Vienna and Monaco), EROS 2000, NIOZ, CEMO (both Netherlands), Musée Océanographique (Monaco) and various laboratories in Europe, Japan and North and South America, to supply recent literature, some of them in press, was highly appreciated. Support for J. Carroll at the IAEA Marine Environment Laboratory was provided by the US Government. Additional thanks are warranted for the US Department of Energy Grand Junction Projects Office and RUST Geotech Inc and to Professors I. Lerche and C.W.S. Moor, Geological Sciences, University of South Carolina USA. The IAEA-MEL operates under an agreement between IAEA and the Government of the Principality of Monaco. The educational value of the book is strengthened by a supplementary demo, called COSMO-BIO, which was kindly made available by its developers, the Dutch National Institute for Coastal and Marine Management (RIKZ), The Hague and the research and consulting company Resource Analysis, Delft.

Table of Contents

Permission has been obtained for a number of figures, listed below, for which the authors express their indebtedness.

Table 0.1. List of reproduced figures.

Fig. N°.	Reference	Journal/book
Plate 2.1	Chen et al. (1994)	Neth J Ses Res
2.7	Duursma et al. (1989)	Sci Tot Environm*
2.10A&B	Duursma and Dawson (1972)	Book Elsevier*
3.2A&B	Duursma and Bosch (1970)	Neth J Sea Res
3.9	Aston and Duursma (1973)	Neth J Sea Res
3.10	"	"
3.11	Duursma and Eisma (1973)	"
3.14A	Li et al. (1984a)	Geoochim Cosmochim Acta*
3.15	Duursma and Eisma (1973)	Neth J Sea Res
3.16	"	"
4.1	Duursma and Bosch (1970)	"
4.2	"	"
4.3	"	"
4.4	"	"
4.18	Duursma (1977)	Deep-Sea Res*
5.7	Duursma et al. (1986)	Neth J Sea Res
5.10	"	"
5.11A&B	Duursma et al. (1989)	"
5.12	"	"
5.13	Duursma et al. (1991)	Mar Chem*
5.14	"	"
5.15	"	"
5.16	"	"
5.17	"	"
6.3	Duursma and Ruardij (1989)	"
6.4	"	"
6.6	Vrie and Duursma (1986)	"
6.7-6.11	Carroll et al. (1993)	Geochim Cosmochim Acta*
7.4	Duursma et al. (1984)	Proc CEC Seminar
7.5	"	"
7.6	"	"
7.7	"	"
7.8	"	"
7.10	Martin et al. (1995)	Mar Chem*
8.19	Hamilton et al. (1994)	J Environ Radioact
8.21	Miquel (1996)	Mar Poll Bull
9.1-9.9	Duursma and Boisson (1994)	Oceanologica Acta
10.3	IAEA (1992)	IAEA Source book

* Elsevier Science

1 Introduction

The terrestrial and marine abiotic and biotic compartments, between which processes and exchange take place, can be defined by three non-living systems: air, water, solid material, and a living system of flora and fauna including man. All elements of the periodic system can be found in these compartments in different concentrations. Their concentrations are regulated by chemical properties of the elements themselves and of the involved compartment matrixes.

The distribution of elements between two compartments is determined by their chemical-physical affinity for the compartment matrix(es) and the transfer parameters regulating apparent equilibria. The exchange reactions within and between the compartments are generated by these sometimes fictitious equilibria, which may never be realized. The same principle is also valid for multi-element compounds, such as nutrients, and inorganic and organic contaminants. The elements also tend to be distributed between different compartments on the basis of their physical-chemical properties.

In this context, the distribution between compartments is never an unpredictable phenomenon. There are always chemical or physical forces driving distribution reactions. This is also valid for biological uptake and loss reactions, even when accumulation occurs against concentration gradients.

The main purpose of this monograph is to demonstrate how the compartmental distribution of elements and compounds may or can be studied. This is presented within a context of understanding and predicting transfer, accumulation and loss of substances in biological and non-biological systems with specific emphasis on the marine environment.

Since environmental compartments rarely have homogeneous matrixes, it is futile to apply thermodynamic laws and theoretical physical-chemical models, with their associated number of restrictions. Therefore only basic theories are presented, directed to practical applications.

2 Processes and equilibria between compartments

2.1 Principles

In order to understand concentration changes in compartments and transfer of molecules from one phase (compartment) to another, it is necessary to define these compartments and the processes involved:

Phase or compartment: Unit of space in which molecules have a defined state of freedom (or non-freedom) of movement and reaction.

Processes: Reactions between molecules and transfer of molecules inside or between compartments.

Equilibria: Reactions and transfer processes being counterbalanced by feedback processes.

2.2 Phases or compartments

2.2.1 Water

Although water is one of the most abundant compounds on earth, it has unusual properties, which are anomalic with regard to practically all the other fluids. The main reason for these anomalic properties can be found in the dipolar molecular structure and the Hydrogen Bonds between different H_2O molecules (Degens 1989). These H-bonds or H-bridges have a strength of 1/20 of the O-H molecular bonds, and as a result that (liquid) water consists of 'clusters' of H_2O molecules.

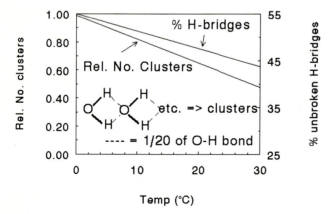

Fig. 2.1. Hydrogen bridges and clusters of water molecules as dependent on temperature.

The % of H-bonds (about 55 % at 0 °C) is temperature dependent (Fig. 2.1).

The capacity of hydrogen atoms to form tetrahedral as well as orthogonal bonding angles allows for a great variety of possible networks or clusters, held together by hydrogen bonds and dipole-dipole interactions.

Thus the liquid state of water, containing a variety of cluster types has to be considered when evaluating the behaviour of elements and compounds dissolved in water. Their behaviour may differ for saline solutions, where the ionic strength of sodium chloride has a marked effect on the internal structure of water clusters. For example, as the ionic strength (salinity) of water increases, the temperature at which water freezes decreases and the temperature at which water reaches its maximum density also decreases. The latter phenomenon is known as the water density anomaly (Fig. 2.2).

Main points can be summarized as:
- The melting point of water is 0 °C. Without H-bonds, the melting point would be -100 °C.
- The boiling point of water is 100 °C. Without H-bonds, the boiling point would be - 80 °C.
- For fresh water the maximum density occurs at 4 °C. For sea water, however, this maximum density is at a lower temperature, depending on salinity (Fig. 2.2).
- The mobility of H^+ in particular, but also OH^- ions is very high in liquid water.

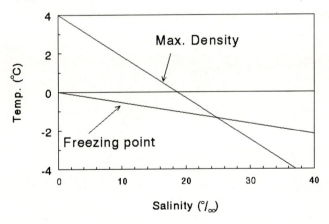

Fig. 2.2. Water density and freezing point anomalies for sea water (Sverdrup et al. 1970; Kennish 1989).

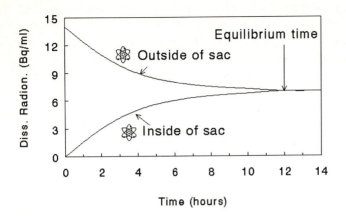

Fig. 2.3A. Empirical dialysis equilibrium times (Dawson and Duursma 1974).

Since water is a polar solvent, vapour or solid, water molecules or its dissociated parts (H$^+$, OH$^-$, ...) are capable of reacting with other polar molecules by forming 'hydrated' molecules or ions. Hydrated ions or molecules have different ionic radii, depending on the extent of hydration, which can be determined by electronic means, or by measuring their diffusion velocities through membranes, cf Fig. 2.3A,B.

The occurrence of substances either dissolved, particulate or colloidal in sea water has to be understood on the basis of the above properties of water.

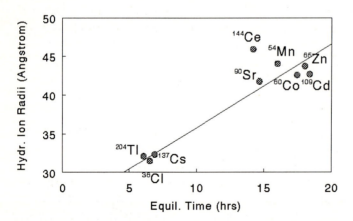

Fig. 2.3B. Ionic radii as related to empirical dialysis equilibrium times as determined with radio tracers. See Fig. 3.5. for setup of experiment.

2.2.2 Particulate matter (PM)

Suspended material is found in practically all natural waters, either floating due to turbulence or a similarity in density to the water or settling through the water column. The lowest particulate matter content in natural waters has been detected by C. Veth, NIOZ, Texel, Nl & co-workers in the 1980s in Antarctic waters on board of the r.v. Polar Stern of the Alfred Wegner Institute, Bremerhaven. They demonstrated that a Secchi disc remained visible to 79 m. This is near the theoretical limit for secchi disc visibility in distilled water.

2.2.2.1 Amounts

The amount (concentration) of suspended particulate matter (PM) in the oceans, coastal seas and estuaries is variable, ranging from a few μg PM/l in open oceans to tens to hundreds of mg/l in estuaries. Some rivers may have loads of several g's/l. A characteristic feature of natural waters is that the PM concentrations can be locally and temporally variable, particularly in estuaries with turbidity maxima. This implies that, for purposes of studying processes in which PM is involved, it is essential to determine PM concentrations as function of time and space.

2.2.2.2 Properties

The essential properties of PM for studying processes of transfer between phases or compartments are related to their (i) Surface-active properties and (ii) Crystal-lattice properties.

(i) Surface-active properties.

Composition
Particulate matter in natural waters consist of biotic debris of living matter, including faeces, organic aggregates (from dissolved organic compounds), atmospheric dust (fine desert materials), minerals (from terrestrial sources) and resuspended bottom sediments. Although the composition and origin of the surface of marine and freshwater PM differs extensively, there are a number of common features with respect to the adsorptive and desorptive properties of particulate matter surfaces which determine the exchange of sorbed (adsorbed and absorbed) and dissolved substances:

- Through adsorption of polar water molecules (intermediate layer), H^+ and OH^- ions on PM, a weak charge is created that allows other ions of opposite charge to attach to PM surfaces.
- Organic matter with surface-active properties can be adsorbed to PM surfaces, producing sites for specific reactions between dissolved substances and the organic molecules.
- In spite of PM having different origins, a number of similarities may be predicted, at least with respect to surface properties.

Specific Surface
Particulate matter, suspended in (sea) water consist generally of low grain-sized solid matter, which has a relatively large specific surface (Sp.S.). The particle surface is in contact with the polar water molecules (and clusters), thus creating a surface double-layer which contains an intermediate layer of 'adsorbed' water clusters.

The ideal grain size-specific surface relation for spheres is given by formula (2.1), valid for spheres:

$$Sp.S. = \frac{4\pi r^2}{\frac{4}{3}\pi r^3 \rho} = \frac{3}{r\rho} cm^2 g^{-1} \qquad (2.1)$$

where r = radius of grains in cm and ρ = specific weight in g/cm³. Using grain sizes (d = 2r) in μm, the specific surface (Sp.S.) is:

$$Sp.S. = \frac{6\cdot10^4}{d\cdot\rho} cm^2 g^{-1} \qquad (2.2)$$

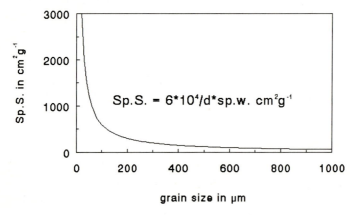

Fig. 2.4. Correlation between the Sp.S. and the grain size (μm) according to formula (2.2).

For example, grains with a size of $1\mu m$ and a ρ of 2.0 gcm^{-3}, the Sp.S. = 3×10^4 cm^{2}g^{-1} (Fig. 2.4).

Determination of surface-active properties
There is no complete theory of the surface-active exchange properties of PM. Natural PM is seldomly composed of ideal spheres. In suspension, sedimentary particles are often present as aggregates due to clustering. These clusters have lower sinking rates as a whole compared to the individual particles making up the cluster, and their size changes (Eisma 1991) due to turbulent water movements (Plate 2.1, Chen et al, 1994).

In order to arrive at relevant measurements, a similar empirical approach is used as for soil science, which encounters similar problems. Effective parameters are: the medium grain-size diameter (D_{50}), the Specific Exchange Capacity (q_t), the Base Exchange Capacity (BEC) and the empirical Specific Surface (Sp.S.).

(a) Medium grain-size diameter D_{50}
The medium grain size D_{50} is operationally defined as the sample grain size whereby 50% of the sediments (by weight) has grain sizes greater than D_{50}, the other 50% (by weight) has grain sizes less than D_{50}. D_{50} is determined by sieving sediments or particulate matter and plotting the % weight of each grain-size fraction against the grain size (Fig. 2.5).

(b) Specific Exchange Capacity, q_t
Another possible relationship can be found in the properties of clay minerals themselves. Each clay mineral has a specific exchange capacity, and by multiplying this value by the relative clay content in a sediment sample, the specific exchange capacity q_t for the sample can be calculated as follows:

$$q_t = \sum\left(\frac{n}{100}\right)\cdot\sum\left(\frac{P}{100}\right)\cdot k \qquad (2.3)$$

where k is 10, 25, 25 and 100 meq/100g for kaolinite, illite, chlorite and montmorillonite, respectively, n = % weight of the size fraction and P is % of mineral in this size fraction (Grim, 1953).

(c) Specific Surface, Sp.S.
The empirical Sp.S. (Specific Surface) is defined as the adsorption of ethyleneglycol (Dyal and Hendricks 1950; cf. also Duursma and Eisma 1973) to sedimentary particles. It is determined by subsequent extraction of the ethyleneglycol and quantitative photometric analysis.

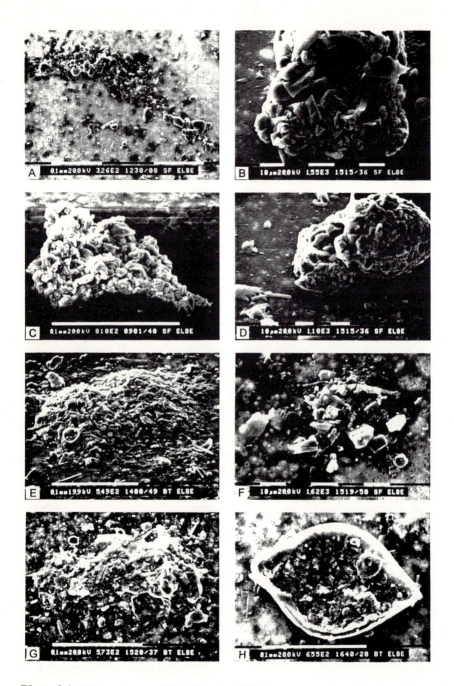

Plate. 2.1. PM as aggregate, from Chen et al. (1994).

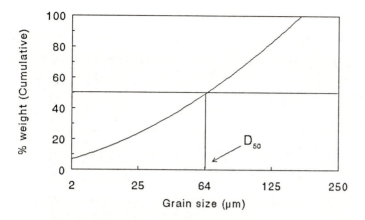

Fig. 2.5. D_{50} determination from a % weight - grain-size plot.

(d) Base Exchange Capacity, BEC.
The Base Exchange Capacity (BEC) is determined by saturating all exchange sites of a sediment or particulate matter sample with Na^+ ions, and replace these subsequently with NH_4^+. The amount of Na^+ in the NH_4Cl-extract is determined by photometric analysis and represents the Base Exchange Capacity in meq/100g (Duursma and Eisma 1973).

(e) Comparison of the empirically determined PM characteristics
In order to evaluate which of the parameters provide the best possible indices for sorptive processes, a comparison has been made of a number of determined parameters (Duursma and Eisma 1973) for 35 ocean and coastal-sea sediments. Table 2.1 summarizes results from the Figs. 2.6 A-H. The parameters **X**: % weight of grain-size fractions, BEC and q_t are correlated to parameters **Y**: D_{50}, BEC, Sp.S., and some other parameters as organic carbon and nitrogen, and leachable Na, K, Mg and Ca.

The most reliable parameter as index of sorptive capacity seems to be the BEC, which is best related to the Sp.S. (R = 0.96), to q_t (R = 0.86), Na and K (R = 0.96 and 0.95, respectively). BEC is less correlated to grain size (R = 0.38; see also exercise 2.1) and organic carbon (R = 0.62). As will be shown later (section 3.2.3.1) for radionuclide sorption, BEC could be used to explain the partitioning between dissolved and particulate nuclides.

Exercise 2.1. In Table 2.1 and Fig. 2.6 are presented a number of correlations of sediment properties. Which of them lack principal causality? Please explain.

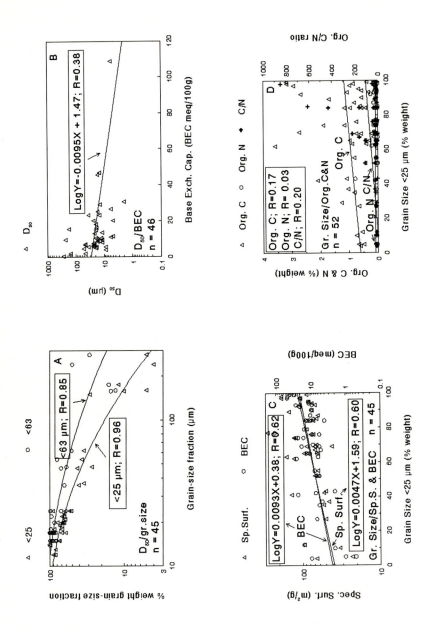

Fig. 2.6. A-H: Curve-fit correlation coefficients (R) of PM surface-active properties with respect to sorptive capacities as summarized in Table 2.1.

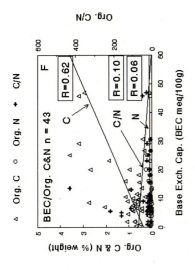

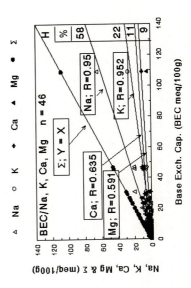

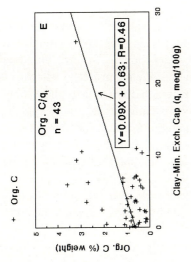

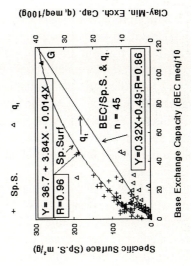

Table 2.1. Summary of curve-fit correlation coefficients (R) of correlations presented in Fig. 2.6. A-H.

No.	Parameter-X	Parameter-Y	Curve-fit (R)
A	% weight Gr.Size $<25\mu m$	D_{50} (μm)	0.96
	% " " $<63\mu m$	"	0.85
B	BEC (meq/100g)	D_{50} (μm)	0.38
C	% weight Gr.Size $<25\mu m$	BEC (meq/100g)	0.62
		Spec. Surf. (m^2/g)	0.60
D	% weight Gr.Size $<25\mu m$	% Org. C; Org. N;	0.17; 0.03;
		C/N	0.20
E	q_t (meq/100g)	% Org. C	0.46
F	BEC (meq/100g)	% Org. C; Org. N;	0.62; 0.06;
		C/N	0.10
G	BEC (meq/100g)	Sp. Surf. (m^2/g);	0.96;
		q_t (meq/100g)	0.86
H	BEC (meq/100g)	Na; K; (meq/100g)	0.96; 0.95;
		Ca; Mg "	0.63; 0.59

Leaching

The exchange properties of metals in sediments are also often investigated by the determination of their capacity of leaching. This property is intended to determine the fraction of metal being 'loosely' bound. Several empirical methods are available, mainly coming from soil science practices where it is necessary to understand the availability of elements from soil to plant roots.

Leaching of metals from marine sediments is usually carried out by applying two media (in excess) of Acetic Acid/Ammonia Acetate and Acetic Acid, creating pH's of 5.4 and 2.3, respectively. The treatment at pH 5.4 leaches the adsorbed part, while leaching at pH 2.3 removes the part that is absorbed, but still can exchange.

An example of such a leaching measurement is given for radiotracers of metals which were previously sorbed by Mediterranean fine-grained clay sediment for 7 months (Table 2.2).

Most peculiar is that Cd and Sr seem to be only adsorbed and not present inside the crystal matrix (absorbed), while the alkaline metal Cs is rather firmly bound. This is due to the fact that Cs is interlayerly bound by a mineral such as illite, since its ionic radius matches the interlayer space of the mineral (Duursma and Eisma 1973).

Table 2.2. Percentage of radiotracer leached at pH 5.4 and 2.3.

Tracers:	Mn	Fe	Co	Zn	Sr	Ru	Ag	Cd	Cs	Ce	Pb
pH: 5.4	50	0	33	4-33	100	10	0	100	2	50	50
pH: 2.3	100	75	100	10-80	100	25	0	100	30	90	50

Organic matter

Particulate organic substances are thought to play a specific role on sorption reactions on particulate matter. They might provide sorption sites, either due to their chemical structure or because of their polar (or non-polar) character. Thus, organic matter, when having a major non-polar character, can inhibit sorption of ionic metal species, while favouring sorption of (in majority) non-polar solutes like organic contaminants. However, since the quantity of particulate organic carbon (POC) attached to the surface, depends also on the available surface, the sorption correlations between POC and sorbed substances may be found to be artificial and thus non-causal. For estuaries it has been determined that there can indeed exist a correlation between POC and the silt content (cf. Fig. 2.7) but the correlation is questionable for POC measured for ocean and coastal sediments, as demonstrated in Fig. 2.6.F & C.

This is confirmed by Milliman (1994) who determined a coupling between the % silt + clay content for organic nitrogen of US Atlantic continental margin sediments, indicating only that sandy sediments contain less organic nitrogen than sediments richer in silt and clay.

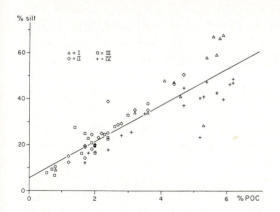

Fig. 2.7. Correlation of % POC and % silt of surface bottom sediments of estuaries of the south-west Netherlands. For the location of stations I, II, III and IV see Fig. 5.11. The correlation coefficient R = 0.88 (Duursma et al, 1989).

Nevertheless correlations found in one region may not be applicable to other regions because of different trends in production, transport and/or accumulation of organic matter. Predicting organic matter from grain size or *vice versa* is therefore highly questionable. This has to be taken into account when sorption processes are studied.

Conclusion: It is essential to determine a complete and quantitative picture of surface-active sites to which compounds can be sorbed. Even when only a few parameters have been determined, the sorption correlations found with these parameters might be non-causal and therefore are potentially misleading. This picture is even more complicated when particulate matter or sediment samples are obtained from a great variety of oceanic regions. The best empirical correlations are between Sp.S., BEC, and subsequently clay-mineral content - which is proportional to Al, Ti - and with some restrictions POC content.

(ii) Crystal-lattice properties

Processes occur between dissolved compounds and the solid crystalline phase of PM. These are usually very slow, and reaching equilibria depends on grain size.

Two kind of absorption (sorption into solid particles) can be distinguished: (a) The migration into (and out of) clay minerals. The solid phase is layered, having possibilities of interlayer transport of compounds. This kind of migration can be selective, since the radii of ions or molecules (partly including their water sphere) should match the distances between the layers.

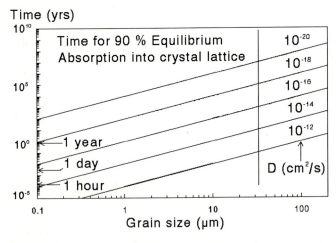

Fig. 2.8. Diffusion front of 90% of original concentration into particles as related to grain size, for different diffusion coefficients (D).

An example is Cesium, whose ions migrate better than those of other alkali metals in clay minerals.

(b) Binding by the solid crystalline phase, other than adsorption to the surface. Also here migration, which might be considered as diffusion into the crystals, can be observed at slow rates. The diffusion becomes obvious only over very long time scales, in particular for large particles (Fig. 2.8). Calculations have been made to predict whether at low diffusion coefficients of 10^{-16} to 10^{-20} cm^2s^{-1} migration would occur in ores buried in geological formations. It was confirmed that for periods of 10^9 yr, significant migration is unlikely for a number of metals (cf. Duursma and Hoede 1976).

2.2.3 Bottom sediments

Many of the reactive properties of PM are evident in marine sediments. Their origin is often the same, mainly because PM is the primary source of pelagic bottom sediments. However, bottom sediments as a compartment, have some additional characteristics. Due to compaction they contain varying interstitial water (pore water), and the contact of individual particles with pore water is different than the contact between PM in suspension and water. In order for attached substances to exchange with the overlying water compartment, they must first pass through the water-sediment interface and diffuse through the pore water. At the same time, there is deposition of new material, while horizontal and vertical displacement of sediments also occurs, either by waves and currents, turbidity currents (on and off slopes) and bioperturbation, such as by ingestion of sediments by deposit feeders.
   Strong tidal currents and wave action (when depths are smaller than half the wave length) may cause redistribution of grains. Depending on grain size and specific weight, sandy bottoms prevail in regions with strong currents, while fine-grained silt sediment are found deposited in current-sheltered regions.

2.2.3.1 Amount

Except for steep sub-mountain or shelf slopes, the ocean and coastal bottoms are generally covered by a layer of sediment, up to some km's thick. Sedimentation rates are variable, ranging from a few mm's per 1000 yr in the open ocean up to some cm's per year on shelves close to river mouths (Sverdrup et al. 1970). This results in accumulation of sediment thickness with time, except in regions where resuspension and erosion events are frequent (Duursma et al. 1996).

2.2.3.2 Properties

The properties of bottom sediment particles are:

(i) Surface-active properties.
Bottom sediment properties are in principle the same as for PM. They affect
interactions not only between particles and water, but also between adjacent
particles in the sediment. A major difference between bottom sediments and
suspended PM is that for bottom sediments interactions can occur over much
longer time scales. PM remains suspended over time scales of days to a few
hundred years, whereas sediment deposits may remain buried for many
thousands of years. Moreover, most marine waters contain oxygen. In bottom
sediments, the depth penetration of oxygen from the overlying water
compartment is limited. Below the depth of oxygen penetration, the chemistry
of interstitial waters may change as bacteria utilize other sources of energy
(like sulphate) to breakdown organic matter (Stumm and Morgan 1981). This
has a particular effect on the chemistry of poly-valent metals such as iron and
manganese (Luther 1990).

Since diagenesis processes (Berner 1980) are not within the scope of this
book, these will not be discussed, and we will limit ourselves to only those
processes which concern periods of time of about a few years to perhaps
1000 yr.

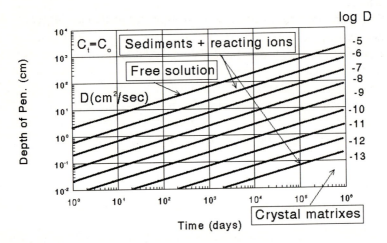

Fig. 2.9. Calculated depth penetration of a 10% concentration front into bottom sediments
(Duursma and Eisma, 1973) from an overlying water mass with constant radionuclide
concentrations, as a function of time for various diffusion coefficients. D is expressed in log
(cm^2/s).

(ii) Bottom sediment properties.
These are also in principle the same as presented for PM. As given in Fig. 2.8, it is possible to present on the basis of a diffusion model the time necessary for a 10% concentration front an element from a constant source will have diffused into a bottom sediment. This demonstrates the dependency of the diffusion on the diffusion coefficient D, which ranges in Fig. 2.9 from 10^{-5} cm^2s^{-1} to 10^{-13} cm^2s^{-1}.

2.2.4 Living organisms

The compartment of living organisms, from which substances are exchanged with other compartments, can be characterized by three major sub-phases. They involve the polar body fluids consisting for more than 90% of water, tissues and the non-polar lipid phase. Although the uptake and loss processes are mainly determined by metabolic activities, food uptake, excretion and exchange through respiratory organs, the final equilibrium between concentrations in organisms and water depends on the affinity of the compounds to these sub-phases.

Thus metals (polar dissolved species) tend to become adsorbed to cell walls and absorbed in body fluids, while non-polar organochlorines will be accumulated in lipids.

In addition of these reactions with the major sub-phases, specific ones are possible, such as:
(a) adsorption to the exterior of organisms, which is proportionally larger for small species such as phytoplankton,
(b) specific binding, e.g. by enzymes or blood cells, and
(c) undefined binding, sometimes through slow uptake rates, such as for methyl mercury in Tuna. Methyl mercury (MeHg) is accumulated in fish with age because it is accumulated with nearly 100% efficiency and practically not released. Inorganic Hg on the other hand is accumulated with about 8% efficiency and released with a biological half-time of about 30 days (Bernhard, 1988).

2.3 Processes and equilibria

The determination of the distribution of natural and anthropogenic chemicals between sea water and the compartments: PM, sediments and organisms has been plagued in the past by conceptual or procedural errors (by either approaching the involved processes too theoretically or by neglecting major competitive reactions that do occur).

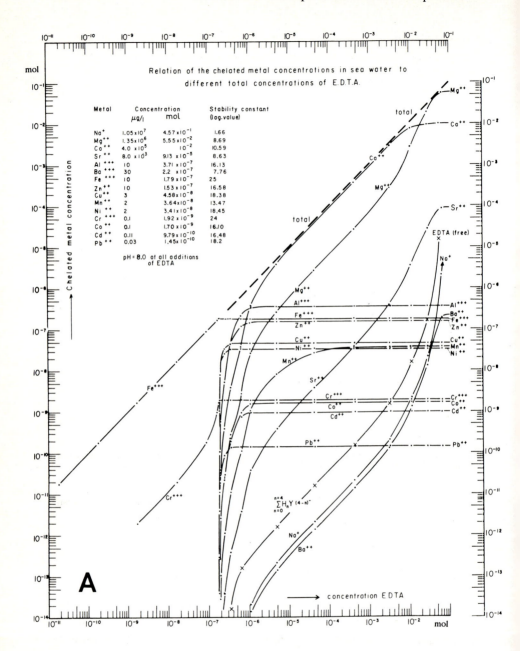

Fig. 2.10A. Relationship of EDTA concentrations with the concentrations of metal-EDTA complexes (MY^-) in sea water as calculated from the total metal concentrations and stability constants, where Y^{4-} = $EDTA^{4-}$. The Fe^{+++} and Cr^{+++} solubility products are not taken into account.

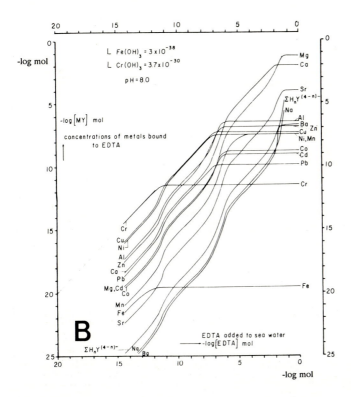

Fig. 2.10B. Idem, but the Fe^{+++} and Cr^{+++} solubility products have been included (Duursma and Dawson 1981).

A number of essential points must be recognized when trying to characterize partitioning of constituents within and between compartments.

2.3.1 Processes/equilibria in the liquid sea water compartment

Sea water and fresh water contain a variety of chemical forms, e.g. cations, anions and uncharged molecules. A fundamental premise of understanding processes and equilibria is that a reaction among any two chemical forms in the dissolved phase will have to compete with reactions that are possible with other (competitive) species. An example will make this clear of the complexation of major metals and some trace metals in sea water with the organic complexing agent EDTA. EDTA, given as the ligand Y^{4-}, can bind metals like Me^{++} into MeY^- (Duursma 1970).

$$Me^{++} + Y^{4-} \overset{K_{st}}{\rightleftharpoons} MeY^{--} \qquad (2.4)$$

where K_{st} = Stability constant, given by:

$$K_{st} = \frac{[MeY^{--}]}{[Me^{++}] \cdot [Y^{4-}]} \qquad (2.5)$$

This stability constant (K_{st}) is similar to a thermodynamic equilibrium constant, based on free molecular movements in the water compartment (Lide 1993).

The relationship between known average metal concentrations in sea water and their stability constants (K_{st}) with EDTA, demonstrates that complexation clearly depends on the concentrations of the various metals present and their stability constants (Fig. 2.10A and B).

2.3.2 Processes/equilibria between water and particulate matter compartments

Despite the above mentioned K_{st} being characterized as a thermodynamic constant, it can not be used as the sole determinant of processes and equilibria occurring between dissolved and particulate species. The required freedom of movement of dissolved substances is different from those of sorbed substances. Nevertheless, partition equilibria will be established between dissolved and sorbed substances at the moment the sorption and desorption rates equalize. This leads to an empirical partitioning coefficient K_d, which is well known in geochemical studies under different names, such as distribution coefficient and concentration factor (Leo et al. 1971).

2.3.2.1 Definition of K_d

2.3.2.1.1 Adsorption to a PM surface layer

The basic partitioning reaction of substances between water and PM is given by:

$$Me^{n+}Cl_n + nNa-PM \underset{k_2}{\overset{k_1}{\rightleftharpoons}} Me-PM + nNaCl \qquad (2.6)$$

in which Me = metal or radionuclide, occurring in concentrations < < < than those of sodium (Na); k_1 and k_2 are the two reaction rate constants, given by:

$$k_2 = f_s \cdot [Me-PM] \cdot [NaCl]^n \tag{2.8}$$

$$k_1 = f_1 \cdot [Me^{n+}Cl_n] \cdot [Na-PM]^n \tag{2.7}$$

At equilibrium:

$$k_1 = k_2 \Rightarrow \frac{f_1}{f_2} = \frac{[Me-PM]}{[Me^{n+}Cl_n]} \cdot \frac{[NaCl]^n}{[Na-PM]^n} \tag{2.9}$$

The second term is quasi constant because the dissolved Na and particulate Na are in excess to the dissolved Me and particulate Me, respectively. As a consequence, changes in dissolved Me to particulate Me will not significantly change the ratio of dissolved Na to particulate Na. Hence the second ratio remains constant. Since also f_1/f_2 is constant at equilibrium, the first ratio [Me-PM]/[$Me^{n+}Cl^n$] also will be constant. This is called the distribution coefficient K_d. Hence:

$$K_d = \frac{[\mu g\ Me/g\ PM]}{[\mu g\ Me/g\ water]} \tag{2.10}$$

At this point K_d is a dimensionless factor of $(\mu g/g)/(\mu g/g)$, which itself is dimensionless on the basis of g/g. Another possibility is to express K_ds as follows:

$$K_d = \frac{[\mu g\ Me/ml\ PM]}{[\mu g\ Me/ml\ water]} \tag{2.11}$$

where K_d is also dimensionless but with ml/ml. There is a constant ratio between these two forms of K_ds:

$$K_d\ (g/g) = \rho \cdot K_d\ (ml/ml) \tag{2.12}$$

where ρ is the specific weight of dry PM. The specific weight of water (loaded with PM) is taken as 1.0 g/ml.

The choice for one or the other incarnation depends on the process of interest. In the case of transport of Me in a river or estuary, where Me is exchanged between PM and water at the same time, the K_d (g/g) has to be used, since the dimension of an exchange process is dependent on weight and time. When however, diffusion in pore waters of bottom sediments are studied, the other K_d (ml/ml) should be used, since the dimension of diffusion is space and time.

Just as the competition exists between metals for complexation in dissolved form, the same problem exists for processes and equilibria of reactions between dissolved and particulate substances. There might be competition between the dissolved complexing (organic and inorganic) ligands and particulate binding (adsorption).

K_ds therefore, have to be understood *only* as apparent distribution or partitioning constants. As long as their values are known in space and time, processes of transfer and equilibrium can be studied using of this factor.

For dilute conditions, with low chemical concentrations in solution, the K_d might approach the sorption/dissolved concentration ratios of the Langmuir and Freundlich isotherm (Fig. 2.11).

2.3.2.1.2 Adsorption to a surface boundary layer of definite thickness

The active surface exchange layer of sedimentary particles may consist of a variable but definite boundary layer, which interacts with the surrounding solution resulting in adsorption and desorption.

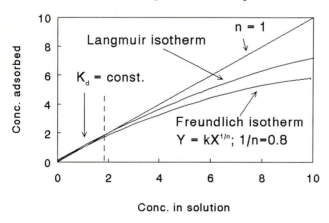

Fig. 2.11. Schematic Langmuir and Freundlich sorption isotherms as compared to a 'constant' K_d. The isotherms are: $KC_{diss} = C_{sorbed}^{(1/n)}$ and $KC_{diss} = (k_1C_{sorbed})/(1 + k_2C_{sorbed})$, respectively.

Abril (1996) developed a so-called microscopic theory for describing the kinetics and distribution of radionuclides between water and particulate matter for this boundary layer. In this theory the free external surface (Sp) is able to capture and release ions from aqueous solution and is defined as:

$$Sp = \int_0^\infty N(1,...)dr \; 4\pi r^2 \cdot 10^{-9} \quad (m^2/m^3) \tag{2.13}$$

where N is number of particles per litre and r the radius of particles. The total mass of this surface boundary layer, msl, in mg/l, can be evaluated as:

$$msl = \frac{4}{3}\pi\rho_p \cdot 10^{-12} \left(\int_0^\xi r^3 N(r,...)dr + \int_\xi^\infty [r^3 - (r-\xi)^3]N(r,...)dr \right) \tag{2.14}$$

where ρ_p is the mean density (kg/m^3), and ξ the average thickness of the boundary layer.

Such an approach has its merits to understanding the dependence of K_ds on grain size, since the mass of the boundary layer becomes relatively larger in relation to the total mass of the particles for smaller grain sizes. Supposing the distribution coefficient K_d, as calculated for the complete particles can be split into a K_{dbl} (K_d of boundary layer) and K_{dcrys} (K_d of crystal lattice), the ratio of K_d to K_{dbl} becomes equation 2.15, in which Q is the μg of sorbed substance for the complete particles and A.Q of the boundary layer, where A is a constant.

$$\frac{K_d}{K_{dbl}} = \frac{\dfrac{Q}{\frac{4}{3}\pi\rho N r^3}}{\dfrac{A \cdot Q}{\frac{4}{3}\pi\rho N\left(r^3 - (r-\xi)^3\right)}} = \frac{1}{A} \cdot \frac{3r^2\xi - 3r\xi^2 + \xi^3}{r^3} \tag{2.15}$$

By taking the ratio ξ/r as α, the equation 2.15 becomes:

$$\frac{A \cdot K_d}{K_{dbl}} = 3\alpha - 3\alpha^2 + \alpha^3 \tag{2.16}$$

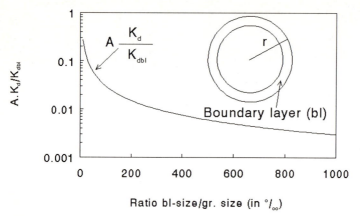

Fig. 2.12A. Curve based on equation (2.16).

Such a correlation given in equation (2.15) produces the relationship presented in Fig. 2.12A. For smaller values of ξ/r, which occur for smaller grain sizes, the $A(K_d/K_{dbl})$ approaches 1, which means K_d approaches $(1/A)K_{dbl}$. Supposing K_{dbl}/A is almost constant, K_d is highest for smallest particles (of identical composition).

The theoretical relationship, however, may not be valid in nature, since the composition of some PM, i.e. clay-minerals is not *a priori* the same for all grain-size fractions (Fig. 2.12B).

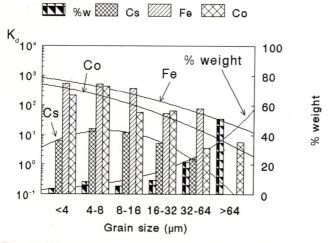

Fig. 2.12B. Distribution of sorbed radionuclides in different grain-size fractions of Dutch Wadden Sea sediment (Duursma and Eisma 1973).

2.3.2.2 K_ds (g/g) and the % Me in solution or sorbed

K_d (g/g) values can be used to determine the % distribution of metals between water and particulate matter. This is an essential conversion since the result will effectively demonstrate how much metal is in solution and how much is bound to particulate matter. Knowing the K_d (g/g), suspended load, water transport and slower PM transport, the effective transport of dissolved and particulate metal can be calculated. It is only necessary to know at each location and for each time what % of the metal is in solution and what % is particulate.

The calculation of % metal dissolved (and particulate) from K_d's and the amount of suspended matter is made as follows using cadmium as an example. Suppose we have: P μg Cd/l (total Cd/solution + PM); S mg PM/l; 1 litre water = 1000 g in which the S mg PM can be neglected; X % Cd in solution, then:

$$K_d = \frac{\left(\dfrac{(100-X)}{100} \cdot P\right) : \left(\dfrac{S}{1000}\right)}{\left(\dfrac{X}{100} \cdot P\right) : 1000} \tag{2.17}$$

which results in:

$$X = \frac{10^8}{SK_d + 10^6} \tag{2.18}$$

Equation (2.18) can be plotted (Fig. 2.13A & B), demonstrating, that for rapidly exchanging metals on PM, the % dissolved can easily be determined.

> **Exercise 2.2.** Calculate (without using formula (2.18) the K_d (g/g) for an estuary containing 20 mg PM/l, where 40 % of Cu is solution; 1 l water contains in total (particulate + dissolved) Q μg Cu.

The answer to this exercise highlights the often used misjudgment of amounts of contaminants in solution and attached to particulate matter. High K_ds indicate that the contaminant (metal, radionuclide or organochlorines) tends to occur in particulate form, but *only* when there is sufficient PM in suspension. This point has been often overlooked or not sufficiently taken into account in the literature on this topic.

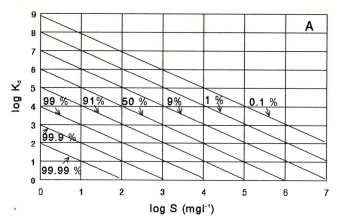

Fig. 2.13A. Percentage of substance in solution for different K_ds (g/g) and quantities of PM in suspension (Duursma and Bewers 1986).

An example to emphasize the value of data on suspended matter concentrations to K_d determinations is that of plutonium, being discharged from the reprocessing plants of Sellafield (Irish Sea) and La Hague (English Channel). Contamination was found at large distances (Aarkrog 1994) from the source (section 7.1 and chapter 8), suggesting that transport was likely through dissolved Pu, even though the K_d for Pu is 10^4-10^6.

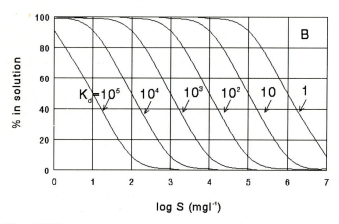

Fig. 2.13B: Percentage of substance in solution for different K_ds (g/g) and quantities of PM in suspension.

Exercise 2.3. ^{239}Pu has a K_d of approximately 10^4 to 10^5 and shows a dynamic equilibrium with sediments and particulate matter (reasonably rapid sorption and desorption). Determine (without using formula 2.18) the % dissolved Pu for a K_d of 5×10^4 in sea water that has a suspended matter load of $10\mu g$ PM/l.

2.3.2.3 K_d (ml/ml) and diffusion

For studies where diffusion is fundamental, the K_d should be only calculated on the dimensionless basis of ml/ml (equation 2.11). This K_d (ml/ml), as we will see later in section 4.1.2 can be applied for the determination of the apparent diffusion coefficient of dissolved substances in pore waters of bottom sediments, where:

$$D_{apparent} = \frac{D_{chloride}}{1 + K_d} \tag{2.19}$$

where the diffusion coefficient of chloride includes the correction due to the matrix of the pore water (porosity etc., see section 4.3).

2.3.3 Processes/equilibria between water and living compartments

2.3.3.1 Uptake and loss models

The fate of chemicals taken up by and released from an aquatic organism can be visualized as passing through different steps related to the size of the organism. For small organisms, from microorganisms to small plankton species, the sorption and excretion acts similarly to the processes occurring with particulate matter. Small organisms can be defined as a two sub-phase system of body fluids and tissue, with uptake and loss occurring through the cell walls. In this case, uptake and loss are a two-step process (Fig. 2.14), and each step exhibits a different transfer coefficient. If these transfer coefficients are taken as diffusion coefficients D_1 and D_2, the model of diffusion in a two-layered sphere can be applied (Duursma and Hoede 1967).
    For small values of time ($t << c^2/D_2\pi^2$), M(t) is:

$$M(t) \approx 8\pi Nb^2 \sqrt{\frac{D_1 t}{\pi}} \tag{2.20}$$

In the case of uptake from water, there first will be an equilibrium-seeking uptake and loss between the chemical dissolved in water and in blood, followed by another equilibrium-seeking uptake and loss process between blood and tissue or lipids.

For intake through ingestion, the equilibria of the substance blood/water and tissue & lipids/blood are in principle the same, but the result may be complicated by the fact that there is a loss from blood to water at the same time as uptake from blood to tissue and lipids.

A similar theoretical approach is given by Pentreath (1973), which was developed for assimilation of radionuclides by fish from water and food. Assimilation, neglecting decay for radionuclides may be described by (2.24):

$$C_t = C_{ss}\left(1 - e^{-Kt}\right) \qquad\qquad (2.24)$$

where C_t = concentration of nuclide in fish at time t, C_{ss} = asymptotic or steady state concentration and K = excretion rate factor equal to $(0.693)/t_{b\frac{1}{2}}$, $t_{b\frac{1}{2}}$ being the biological half time.

When uptake balances excretion (2.25):

$$\frac{dC_t}{dt} = KC_{ss} - KC_t = 0 \qquad\qquad (2.25)$$

Equation (2.25) can be modified for body size, rate of growth, temperature and input from water and food. For intake from water this is expressed as (2.26):

$$C_t = C_0\, e^{-(K_T+\lambda_g)t} + \frac{Iw_t}{(K_T+\lambda_g)}\left[1-e^{-(K_T+\lambda_g)t}\right] \qquad\qquad (2.26)$$

where C_0 = concentration in fish at t = 0, Iw_t = intake from sea water/day, λ_g = growth constant/day and K_T = excretion/day. In controlled aquarium conditions at steady state, formula (2.26) gives (2.27):

$$C_t = \frac{Iw_t}{(K_T+\lambda_g)}\left[1 - e^{-(K_T+\lambda_g)t}\right] \qquad\qquad (2.27)$$

and, when $C_1 = C_{ss}$ and $t = \infty$, this gives:

$$Iw_t = C_{ss} \left(K_T + \lambda_g \right) \tag{2.28}$$

If concentrations in the aquarium can be held at a constant level of activity/unit weight of water (allowing for decay, which gives λ_g is zero), the uptake of nuclide can be expressed in units of water/g/t. This allows quantitative labelling of water, using inactive food, or labelling the food, or a combination of both.

When the concentration factor CF (of K_d) is defined as

$$CF = \frac{[nuclide]_{fish}}{[nuclide]_{water}} \tag{2.29}$$

then the input can be expressed in nuclide/g/day in one ml sea water to maintain C_{ss} (Table 2.3).

Table 2.3. Input of radionuclides/g fish/day in units of water, assuming an f value of 1, to maintain (C_{ss}), according to Pentreath, 1973. C_{ss} = steady state concentration/g in ml sea water.

$t_{b\frac{1}{2}}$ (days)	C_{ss}			
	10^1	10^2	10^3	10^4
10^1	0.69	6.9	69	690
10^2	0.069	0.69	6.9	69
10^3	0.0069	0.069	0.69	6.9

Another model is given by Sijm et al. (1992), based on the description given in Fig. 2.16A and B. In their model, the kinetics of the amount of chemical entering and leaving the fish can be described by (2.30):

$$\frac{dn}{dt} = k_1 W C_w + E F C_{fd} - (k_2 + k_m) n - R k_r n \tag{2.30}$$

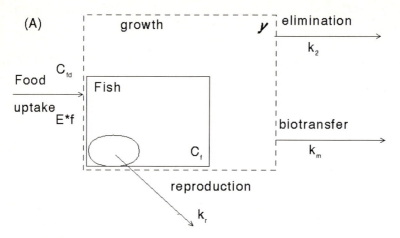

Fig. 2.16A. Biomagnification kinetics of hydrophobic chemicals for fish, where C_{fd} = concentration in food, C_f = concentration in fish, E = uptake efficiency, f = feeding rate, γ = growth rate constant, k_2 = elimination rate constant, k_m = biotransformation rate constant, k_r = elimination rate constant for reproduction rate.

where n= Σ[chemical] in organism (mg), t = time (week), k_1 = uptake rate constant [(liter).(kg of fish)$^{-1}$.week^{-1}], W = weight of fish (kg), C_w = [chemical] in water (mg.l^{-1}).

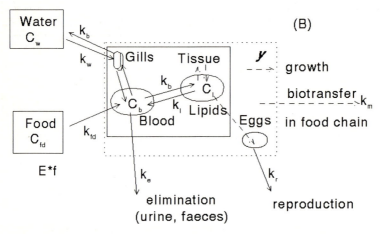

Fig. 2.16B. More detailed view, but with similar kinetics as is shown in 2.16A. E is absorption efficiency of chemical from food (between 0 and 1), F = food consumption [(kg food).week^{-1}], C_{fd} = conc. chemical in food [mg.(kg food)$^{-1}$], k_2 = elimination rate constant [week^{-1}], k_m = biotransformation rate of the chemical [week^{-1}], k_r = zero-order reproduction rate and R is a trigger value (either 1 or 0, depending whether reproduction takes place or not, respectively).

Using the described model, Sijm et al. (1992) determined a range of elimination rates of PCBs (polychlorinated biphenyls) from fish (rainbow trout, guppy and goldfish). Values of $k_2 + k_m$ varied between 0.0006 and 0.03 day^{-1} for muscle and whole fish. Taking growth (γ) into account, changed the estimates to $k_1 + k_m + \gamma = 0.010 - 0.147$ week^{-1}.

Under steady state conditions, dn/dt of formula (2.30) = 0, and the ratio of the chemical in the fish relative to the food can be determined. This ratio is given as the biomagnification factor K_m in formula (2.31):

$$K_m = \frac{Ef}{(k_2 + k_m + \gamma)} \tag{2.31}$$

where γ = is growth rate constant (week)$^{-1}$ from $W_t = W_0 e^{\gamma t}$, in general $\gamma <$ 1, f = feeding rate factor from $F_t = fW_t$, where F = feeding rate [(kg food).(kg fish)$^{-1}$.(week)$^{-1}$]. K_m ranges between 0.03-6 (including growth γ) and 0.04-77 (exclusive growth γ), depending on the kind of PCB congeners studied (Table 2.4).

Table 2.4. Biomagnification factors (K_m) of PCB congeners and two groups of Guppy, Group I: during 30 weeks of exposure. Group II: during 120 weeks of exposure. For C_f/C_{fd} t = 30 weeks. Feeding rates are 0.25 and 0.22 kg food/kg fish for group I and II, respectively (Sijm et al. 1992).

IUPAC no.	C_f/C_{fd}	K_m (+ growth)	K_m (no growth)
		Group I	
52	0.80	1.04	2.5
80	0.89	1.13	2.3
153	1.37	2.84	43
195	1.29	2.50	24
209	0.89	2.28	>60
		Group II	
52	1.19	1.60	3.3
80	0.035	0.035	0.043
153	0.045	0.033	0.042
195	0.66	1.19	4.3
209	1.38	5.94	>77

The question may be posed whether such a biomagnification factor (K_m) has a correlation with an equilibrium partition factor (or partition coefficient)

between PCB concentrations in water and in fish, upon digestion of non-
contaminated food or starvation. There seems to be no clear answer at this
point. Probably, as discussed above, there will be a tendency for a first
equilibrium between PCBs in blood and water and a second equilibrium
between PCBs in lipid and in blood.

A confounding factor is that feeding seems to be essential for maintaining an
active metabolism process, as will later be discussed in section 5.3.3.5. for
PCBs in fasting eel. This is also confirmed for accumulation, retention and loss
of the radionuclide ^{95m}Tc by the edible winkle (*Littorina littorea* L.), as
investigated by Swift (1989). The accumulation of ^{95m}TC from sea water has
a good fit to an equation of the form (2.32):

$$C_t = C_{ss}\left(1 - e^{-kt}\right) \tag{2.32}$$

where equilibrium is reached after t = 210 days, C_{ss}/C_t = 45 and k = 0.006.

Retention after this period of 210 days proceeded with different rates for
winkles being fed and winkles kept under starvation conditions (Fig. 2.17).
Then the elimination of ^{95m}TC was describable by:

$$
\begin{aligned}
A_t &= 100 \cdot e^{-0.002t} \quad (\text{starved}) \\
A_t &= 96 \cdot e^{-0.018t} \quad (\text{fed})
\end{aligned}
\tag{2.33}
$$

The biological half-time for starved winkles ($t_{b\frac{1}{2}}$ = 347 days) exceeded far
that of fed winkles ($t_{b\frac{1}{2}}$ = 39 days). This demonstrates the dependency of
excretory processes and ultimate contaminant clearance rates on active
metabolism.

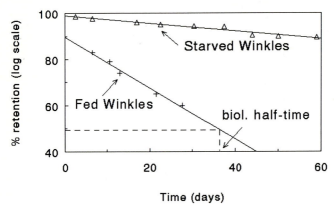

Fig. 2.17. Retention of ^{95m}TC by fed and starved winkles (after Swift 1989).

2.3.3.2. Barriers to uptake

Potential barriers to uptake across the gill are determined by a number of factors among which are the physico- and biochemical properties of the substance (Erickson and McKim 1990), such as molecular size, lipophilicity, binding to blood proteins and affinity to the gill epithelium. In fact Hayton and Barron (1990) distinguish four rate-limiting barriers, (a) transport by water flow to close proximity to the gill epithelium, (b) diffusion across the aqueous stagnant layer to the epithelial surface, (c) diffusion across the epithelium and d) distribution throughout the body of the animal by the blood (Fig. 2.18).

Table 2.5. Symbols in the gill uptake model.

Symbol	Units	Definition
A	cm^2	Epithelial absorbing surface area
D_m	$cm^2.t^{-1}$ (t=time)	Diffusion coefficient in epithelium
K_m	unitless	Epithelium/water distribution coefficient
d	cm	Thickness of epithelium
D_a	$cm^2.t^{-1}$	Diffusion coefficient in water
h	cm	Thickness of aqueous stagnant water
K_b	unitless	Blood/water distribution coefficient
V_b	$cm^3.t^{-1}$	Effective blood flow through gill
V_w	$cm^3.t^{-1}$	Effective water flow past gill
P	$cm^3.t^{-1}$	Uptake clearance

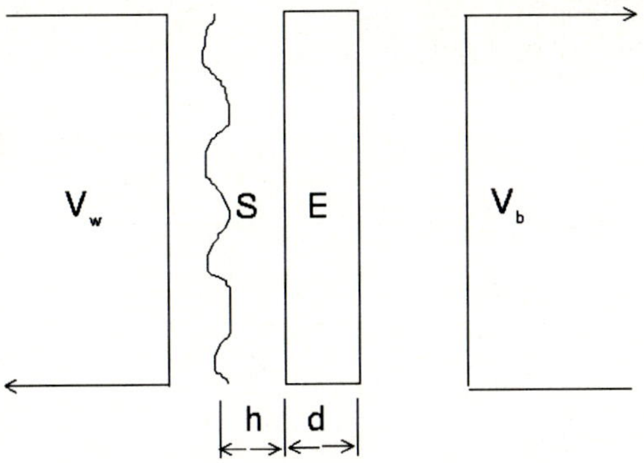

Fig. 2.18. Schematic representation of the gill and rate-limiting barriers to contaminant uptake. For the symbols see Table 2.5.

The overall resistance, being the sum of the individual resistances is described by (2.34):

$$ P = \left[\frac{d}{D_m \cdot A \cdot K_m} + \frac{h}{D_a \cdot A} + \frac{1}{K_b \cdot V_b} + \frac{1}{V_m} \right]^{-1} \qquad (2.34) $$

Hayton and Barron (1990) give some values for gills (goldfish). For small molecules a value of $D_a = 1 \times 10^{-5}$ cm^{2}s^{-1}, giving a retention value for the stagnant layer (second term of equation 2.34) of about 2.5 cm^3.min^{-1}/g body weight. This exceeds that of the blood limited uptake (third term of equation 2.34), which is about 0.04 cm^3.min^{-1}/g total body weight, in case $K_b = 1$.

3 Complexing metal- and radionuclide-sediment reactions

Theories on complexation reactions in aquatic suspensions are difficult to model. The thermodynamic conditions of free movement of molecules in solution do not fully exist for molecules attached to particulate matter. Therefore, in order to understand marine systems, a more or less stepwise approach is presented to demonstrate empirically shortcoming of current theories. Using examples from a number of experiments, results are presented which demonstrate complexation, sorption and diffusion reactions which occur in the dissolved and sedimentary compartments.

3.1 Complexation of metals by dissolved organic matter

The complexation of metals and radionuclides by dissolved organic matter or inorganic species is a process that competes with sorption of these same metals to suspended particulate matter (PM) and bottom sediments. This complexation plays a role in determining whether contaminants remain to a greater extent in solution and follow a different transport pattern than those attached to suspended matter and sediments. The consequences of complexation versus adsorption of contaminants can best be demonstrated by an experiment indicating the kind or problems involved and possibilities to match results with model calculations.

3.1.1 Experiments of competition between complexation and adsorption

Compartments: Marine (Mediterranean) clay sediment (S), Sea water containing dissolved radioisotopes ^{60}Co and ^{65}Zn (Me^{++}) and leucine as chelating dissolved ligand (L^-).

Reactions: Complexation in the dissolved compartments shown by:

$$Me^{++} + L^- \rightleftharpoons MeL^+ \quad K_1 = \frac{[MeL^+]}{[Me^{++}][L^-]} \tag{3.1}$$

$$MeL^+ + L^- \rightleftharpoons MeL_2 \qquad K_2 = \frac{[MeL_2]}{[MeL^+][L^-]} \tag{3.2}$$

Sorption (adsorption mainly) from the dissolved to the particulate compartment is given by:

$$Me^{++} + S \rightleftharpoons MeS \qquad K_d = \frac{[MeS]}{[Me^{++}]} \tag{3.3}$$

Experiment: A suspension containing PM (here S), ^{60}Co, ^{65}Zn and leucine in sea water is homogenized. Sampling is carried out by removing 5 ml of suspension at increasing time intervals. Subsequent filtration produces a filtrate and PM on a filter. These are analyzed for radioactivity/ml filtrate and radioactivity/g PM. The partition coefficient K′, thus determined, becomes:

$$K' = \frac{[MeS]}{[Me^{++}+MeL^++MeL_2]} \tag{3.4}$$

From the equations 3.1, 3.2, 3.3 and 3.4 it is possible to express the K′ (measured as the 'apparent' K_d of the experiment) as a function of K_1, K_2, K_d and L′, being:

$$\frac{1}{K'} = \frac{1}{K_d}\left(1 + K_1[L^-] + K_1K_2[L^-]^2\right) \tag{3.5}$$

where [L′] can be separately calculated from a number of equations in which the total of the various species of metal-ligand and free ligand equals the total added leucine, and the sum of all metal species equals the total amount of metal (here radionuclide, see further Duursma 1970). The results show that the calculated 1/K′ values for ^{60}Co and ^{65}Zn seem to match the determined 1/K′ only when the competition of the major sea water ions H$^+$, Ca^{++} and Mg^{++} is also taken into account (Fig. 3.1). For these calculations the formula 3.1 to 3.5 have to be extended for these cations.

These results suggest that reactions of metals in the marine environment between water and particulate matter compartments depend on the speciation of these metals and their complexing with either dissolved organic (or inorganic) substances. Clearly chelation between trace metals and dissolved organic matter in both the sea and in fresh water is not a simple process. Competition for chelation exist between trace and major metals, including H$^+$, where at low DOC concentrations and low stability constants, the % chelation may become very low.

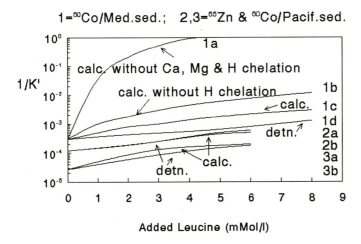

Fig. 3.1. Determined and calculated reciprocals of the apparent distribution coefficient K′ of sorption of Co and Zn by Mediterranean and Pacific sediments, in relation to the amount of leucine added as complexing agent 1: ^{60}Co + Mediterranean sediment; 2: ^{65}Zn + Pacific sediment; 3: ^{60}Co + Pacific sediment.

For a number of cases, this can be approached by theoretical means, however only when equilibria and stability constants are known.

Remark: For students with experience in computer modelling, the above-shown approach can be easily modeled.

3.2 Results on radionuclide sorption by marine sediments

3.2.1 Methodologies of K_d determination

Different techniques exist for determining K_ds (Sioud, 1994), from simply shaking a suspension for a short period, spiked with a chemical, to much more sophisticated ones. The sophistication of the technique employed depends on the purpose for which K_d needs to be determined. Designing the experiment with the purpose in mind is essential, because the time scales of the processes being studied may differ from hours to thousand of years. Furthermore, whether a K_d is a constant or a variable factor, depends on environmental conditions including concentration, temperature, turbidity (PM in suspension) and time. These points should always be kept in mind either for the determination of the K_ds or during their application.

The simplest method for the determination of K_ds is to arrange a suspension and sample the solution and PM at increasing time intervals. The K_d from the determined concentrations according to formula 2.10 and 2.11 is defined as K_d = [Me-part.]/[Me-diss.].

Due to technical difficulties a number of procedural errors are frequently made: (i) the [Me-part.] is determined by difference between the total Me in suspension and that in the filtrate: [Me-part.] = [total Me] - [Me-diss.], (ii) possible precipitation of Me on the walls of the flask is not taken into account, and (iii) the extent of metal concentration and suspended matter quantities are not taken into consideration. Error (i) results from the fact that precision of the determination of [Me-part.] by difference may be very low. Thus, variability in both determinations of [Me-total] and [Me-diss.] may be larger than [Me-part.]. Additionally [Me-total] may decrease with time when there is adsorption of Me to the walls of the experimental vessel. Errors (ii) + (iii) are apparent when [Me-diss.] is high with respect to its solubility and forms precipitates. For example Hg^{++} in the presence of S^-, under anoxic conditions, gives $HgS\downarrow$.

In order to investigate all possible experimental artifacts on the determination of a 'constant' K_d, a number of tests will be described here (also see Duursma and Bosch 1970).

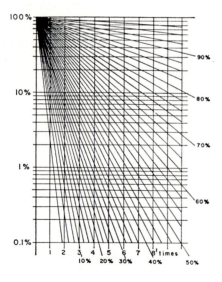

Fig. 3.2A. A reproduction of this graph on tracing paper can be used to find the average % reduction per treatment.

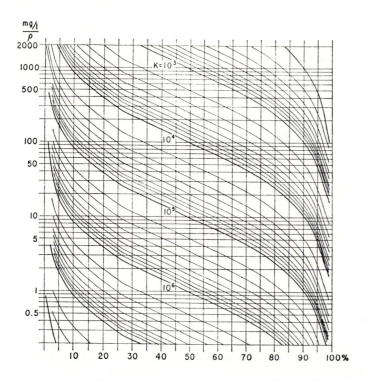

Fig. 3.2B. Theoretical correlation between the % determined in Fig. 3.2A with the amount of PM added per treatment, expressed as mg.l⁻¹.ρ⁻¹, which is dimensionless (Duursma and Bosch 1970).

3.2.1.1 Various techniques of Kd determination

3.2.1.1.1 Settling or sedimentation technique

This technique (Fig. 3.3) was initially developed with the objective of simulating the effect of scavenging of dissolved radionuclides by settling PM in the sea. A practical application is the spraying of sludge over a contaminated area where radionuclides need to be confined to a certain location, i.e. after and accident with a nuclear propulsed vessel, where it is essential to have the released high-level short-living fission nuclides confined at the area in the first period after the accident.

Step 1: A cylinder is filled with (filtered) sea water containing the radiotracer.

Step 2: At increasing time intervals, a 1 ml suspension of PM in sea water is carefully released just below the water surface.

Step 3: One ml water is sampled from the top layer, just before adding the 1 ml PM suspension.

Step 4: K_d is calculated from the [Me-diss.] decrease with time from the equation:

$$\frac{S}{\rho} = \left(\frac{100}{P}-1\right)\cdot\frac{10^6}{K_d} \tag{3.6}$$

where S is the amount of PM added each time, ρ is the specific weight of the PM (dry), P is the % reduction each time and K_d is the distribution coefficient in ml/ml. (If determining K_d in g/g, the factor ρ should be left out.).

The graphs in Figs. 3.2A and 3.2B nay be used to quickly estimate the % uptake.

Remark: A control cylinder filled only with water and tracer is used to quantify the amount of tracer sorbed onto the walls of the vessel.

3.2.1.1.2 The thin-layer technique

The thin-layer technique was developed with the objective of simulating adsorption of dissolved radionuclides onto the very top layer of bottom sediment. Therefore the following procedure was performed (Fig. 3.3):

Step 1: A glass dish is prepared containing i.e. 150 ml filtered sea water, to which a known amount of radiotracer is added.

Step 2: Filter preparation: From a defined sediment suspension, a 1 ml aliquot, containing 10 mg of sediment, is filtered through a 1 inch 0.45 μm Millipore filter. Care is taken that the sediment is homogeneously distributed.

Step 3: Ten filters + sediment are placed in a glass dish, containing the known amount of sea water and tracer. Ten blank filters are also placed as controls.

Step 4: At increasing time intervals 1 ml of water and one of each filters (+ sediment and control) is sampled.

Step 5: After preparation for analysis (i.e. gamma counting), the activities of radionuclides are determined per ml(= g) water and g PM, respectively.

Step 6: K_d is calculated from Fig. 3.4., according to the formula (2.10), where the [Me-part.] is the activity on the filter + sediment minus that of the control, and [Me-diss.] is that of the filtrate.

Control: For understanding confounding reactions, a budget has to be made to calculate the total amount present in solution and sediment for each time t. The difference of this amount with that originally added reflects wall adsorption but does not play a role in the K_d determination.

Three Sorption Techniques

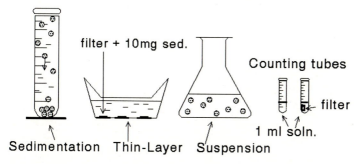

filter + 10mg sed.

Counting tubes

filter

1 ml soln.

Sedimentation   Thin-Layer   Suspension

Fig. 3.3. Scheme of techniques to determine K_dS.

3.2.1.1.3 The suspension technique

This technique simulates partitioning in suspensions between water and particulate matter in the sea.

Step 1: A vessel is prepared with a known amount of sea water and radiotracer. PM is added to obtain a suspension with known amount of mg PM/l.

Step 2: Stirring may occur by different means, from shaking by hand four times a day, to continuously stirring. *Remark:* Magnetic stirring might cause anomalies due to grinding of the PM see Fig. 3.16.

Step 3: 1 ml suspension is sampled and subsequently filtered.

Step 4: Radioactivity is measured of the filtrate and the filter.

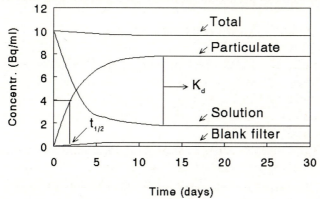

Fig. 3.4. Plot of sorption results to determine K_dS by thin-layer and suspension techniques.

Step 5: K_d is determined as above, where sorption is equal to adsorption + precipitation. For obtaining the filter blank, the filtrate is refiltered. Equally a budget control is made for checking wall adsorption.

3.2.1.1.4 Dialysis technique

In order to compare uptake of radionuclides by phytoplankton and particulate matter, Dawson and Duursma (1974) developed a dialysis technique (Fig. 3.5). The technique is based on the passage of nuclides through a dialysis membrane while keeping the spiked solution separate from either the PM or phytoplankton suspension. Although the passage of these nuclides depend on their ionic radius (including hydration; see Fig. 2.3A,B), equilibrium is attained in periods of hours (Fig. 3.6).

The technique can be used equally by determining competition between sorption of particulate matter and phytoplankton directly. This is done by placing two dialysis sacs in the spiked solution and determining the equilibria concentrations of the three solutions (Fig. 3.5).

Exercise 3.1. Calculate both the $K_{d(PM)}$ and $K_{d(pl)}$ from the results given in Fig. 3.6, by assuming that the suspension contains 100 mg PM/l and the phytoplankton suspensions has 500 μg plankton (dry weight)/l. Express the K_ds in g/g units.

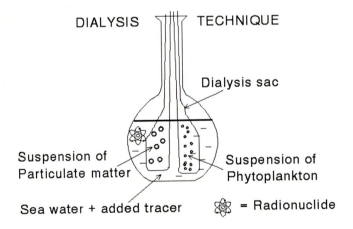

Fig. 3.5. Determination of the partition of spiked radionuclides by the dialysis technique.

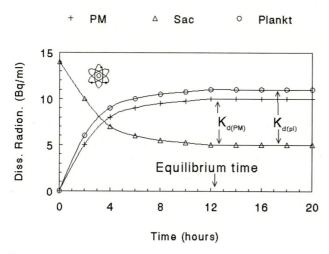

Fig. 3.6. Example of dialysis equilibrium attained in a period of 12 hours.

Results of the last technique demonstrate that for 13 radionuclides, competition exists between the PM and phytoplankton compartments for the spikes added to the water compartment (Fig. 3.7).

In a natural environment, this has major implications since it suggests which route the radionuclide might go when introduced into the water compartment: either through PM to bottom sediments or through phytoplankton into the food chain.

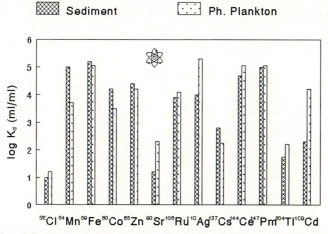

Fig. 3.7. K_ds of various nuclides and one pelagic clay marine sediment for PM and phytoplankton determined with the dialysis technique (Dawson and Duursma, 1974).

At the start, the distribution between PM and phytoplankton compartments can be calculated, taking into account initial PM and phytoplankton concentrations. When all concentrations are normalized to $\mu g/ml$ (Dawson and Duursma 1974), this gives:

$$T = w + s \cdot S + p \cdot P \tag{3.7}$$

where w, s and p are the radionuclide (or trace metal) concentrations in $\mu g/ml$ for water, dry sediment and phytoplankton, respectively, S and P the concentration of PM and phytoplankton in suspension in $\mu g/ml$ (PM solid weight, phytoplankton wet volume). When $K_{d(PM)}$ and $K_{d(pl)}$ are the distribution coefficients of PM and phytoplankton, respectively, the % nuclide (or trace metal) in the three compartments at equilibrium become:

For water:

$$\frac{w \cdot 100\%}{w + wSK_{d(PM)} + wPK_{d(pl)}} = \frac{100\%}{1 + SK_{d(PM)} + PK_{d(pl)}} \tag{3.8}$$

For PM:

$$\frac{wSK_{d(PM)} \cdot 100\%}{w + wSK_{d(PM)} + wPK_{d(pl)}} = \frac{100\%}{1 + \dfrac{1}{SK_{d(PM)}} + \dfrac{PK_{d(pl)}}{SK_{d(PM)}}} \tag{3.9}$$

For phytoplankton:

$$\frac{wSK_{d(PM)} \cdot 100\%}{w + wSK_{d(PM)} + wPK_{d(pl)}} = \frac{100\%}{1 + \dfrac{SK_{d(PM)}}{PK_{d(pl)}} + \dfrac{1}{PK_{d(pl)}}} \tag{3.10}$$

This results in a distribution over the PM and phytoplankton compartments by combining (3.9) and (3.10):

$$\frac{\%\ nuclide(tracemetal)/PM}{\%\ nuclide/Phytoplankton} = \frac{SK_{d(PM)}}{PK_{d(pl)}} \tag{3.11}$$

3.2.1.1.5 Estimation of K_ds of elements of the periodic system in the oceans for modelling health physics aspects of nuclide contamination

K_ds play an essential role in a determination of an oceanographic and radiological basis for the definition of high-level wastes, unsuitable for dumping at sea by the International Atomic Energy Agency, hereafter referred as IAEA (IAEA, 1984; 1985). The determination is based on the likely extent of exchangeable phase element concentrations in sediments and their geological abundance. A comparison is made between the abundances of elements in pelagic clays with those in continental source rocks.

Where an element appears to be enriched in pelagic clays relative to crustal rocks or average shales, enrichment is taken as a measure of the likely authigenic augmentation in particles in their transport through the ocean and to the bottom. Out of 58 elements for which long-term K_ds were required for the IAEA models, 21 gave positive enrichments of the sediments and therefore crude estimates of exchangeable-phase concentrations.

The remaining elements had to be dealt with the in a wholly subjective manner by assigning a 'fixed' proportion of 10% of the total concentration of the elements in pelagic clays as potentially exchangeable.

A compilation of the K_ds thus obtained is given in Fig. 3.8. from the data of IAEA (1984) and schematized by Duursma and Bewers (1986).

Exercise 3.2. Calculate the average ocean exchange K_d of a hypothetical element X, which has an average concentration of 10 $\mu g/l$ in sea water, of 150 $\mu g/g$ in pelagic clay and of 120 $\mu g/g$ in source rock.

3.2.2 Dependency of K_ds on concentrations in solution and amounts of PM in suspension and in consolidated sediments

3.2.2.1 Carrier concentrations and K_ds

Changes in dissolved Me concentrations may affect adsorption, and thus the value of the K_ds. As given in equation (2.9), K_d may be considered solely constant when the concentrations of [Me-diss.] and [Me-part.] are relatively low compared to which the Me is exchanging, i.e. Na^+. The same is expected when the K_d represents the concentration ratios for the Freundlich and Langmuir isotherm at low concentrations (Fig. 2.11). This will be different when either the Na concentrations are low (fresh water) or when radionuclides have high carrier concentrations.

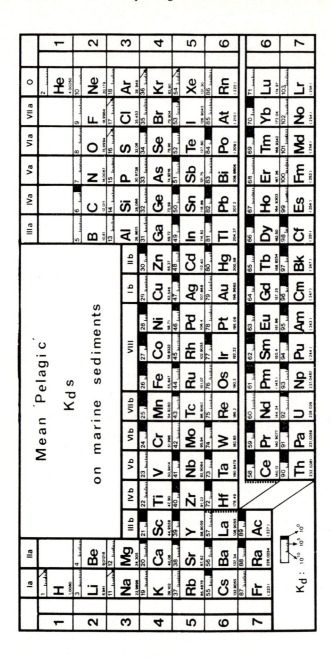

Fig. 3.8. Compilation of recommended mean K_d values (after IAEA, 1984), presented as bars on a log scale, for a number of elements of the periodic system, based on the estimate of element enrichment in pelagic clay enrichment in relation to source rocks. Values of K_d equal to 1 are indicated by an arrow.

Strangely enough, it is very difficult to find clear evidence of how the K_ds should correlate with carrier concentrations.

Correlations seem to be different from one radiotracer (^{137}Cs, [65]Zn, ^{60}Co and [106]Ru) to another and for different ocean sediments (Fig. 3.9). The reason may lie in the kind of sorption reactions, ranging from ion exchange, chemical binding to precipitation, or a combination of reaction types Aston and Duursma 1973).

The conclusion has to be drawn that for K_d determinations under different carrier concentrations, the K_ds have only an empirical value for the conditions under which the determination has been executed. Extrapolation to other metal or radionuclide-carrier concentrations is not possible.

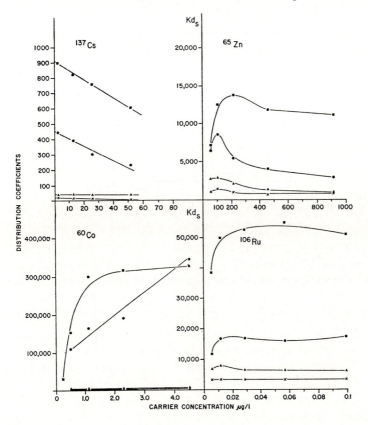

Fig. 3.9. Correlation of K_ds with different carrier concentrations ^{137}Cs, [65]Zn, ^{60}Co and [106]Ru. The sediments used are: x Bahamas; • Mediterranean; ▲ South Pacific; ■ Bombay Harbour.

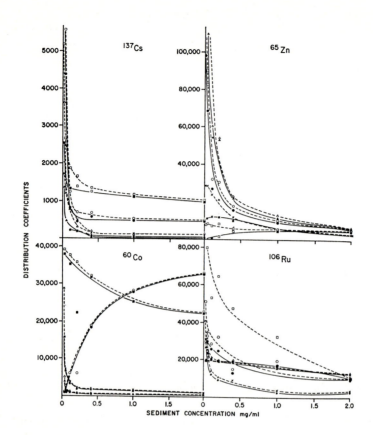

Fig. 3.10. K_ds for ^{137}Cs, ^{65}Zn, ^{60}Co and ^{106}Ru as function of PM concentrations (Aston and Duursma 1973). The symbols represent the same sediments as given in Fig. 3.9. Open symbols K_ds corrected for particulate fraction of the isotopes. See for the characteristics of these sediments (exchange properties, clay minerals, grain-size fractions, etc.) Duursma and Eisma (1973).

Exercise 3.3. Calculate the specific radioactivity of a radionuclide tracer ^{237}Pu ($T_{1/2}$ = 45 days) and ^{239}Pu ($T_{1/2}$ = 24400 years) in Becquerel (Bq)/g and Curie (Ci)/g, where $T_{1/2}$ is half-time of decay, equal to $0.693/\lambda$, λ being the decay constant of $N = N_0 \, e^{-\lambda t}$.

3.2.2.2 PM concentrations and K_ds

The presence of suspended PM will enhance precipitation since PM acts as a nucleus for precipitation. It is difficult to distinguish this process from 'real' sorption during the determination of the K_d. Probably the thin-layer method gives the best opportunity, since the blank filters also act as a site for precipitation. For the two other methods, the suspension and settling techniques, a control (filtering the spike solution) without PM might not give the same result.

Except for a few anomalies, K_ds can be rather constant when suspensions are investigated ranging from 0.3 to 2.0 mg PM/ml (300 mg PM/l - 2 g PM/l) (Fig. 3.10). These PM concentrations are, however, rather high with respect to natural marine situations, where PM concentrations occur from < 10 μg/l to 1000 mg/l.

For lower concentrations (< 0.3 mg PM/ml; Fig. 3.10) K_ds may increase exponentially with lower PM concentrations. This was valid for the nuclides ^{137}Cs, ^{65}Zn, ^{60}Co and ^{106}Ru; the only exception was ^{60}Co for PM (B).

The explanation given was that these nuclides in all cases have some tiny fraction of particulate nuclide. When PM concentrations are very low, this fraction become obvious as a large concentration per g of PM. From formula (2.10) it can be seen that this affects the numerator of the K_d, forcing the result of the calculation to be enormously high.

At large PM concentrations this fraction is negligible, resulting in the conclusion that in confined bottom sediments K_ds may be rather constant.

3.2.3 Correlations of K_ds with sediment and metal properties

There has been intensive debate on how exchange processes occur between metals in solution and available sites on PM. The prevailing view is that only the ionic metal species take part in the adsorption-desorption reactions (Li 1981; Balistrieri and Murray 1984; Nyffeler et al. 1984), despite a lack of experimental verification, except for those mentioned earlier with organic complexation (cf. section 3.1).

There are many kinds of adsorption reactions for particulate matter, as described in section 2.2.2. They include ion exchange, chemical binding and precipitation-dissolution. Also, many scientists believe that there exists a threshold phenomenon; that is, a finite number of exchange sites exist on PM and these sites may become saturated (Benjamin and Leckie 1980; 1981; Balistrieri and Murray 1983).

However, for ion exchange alone, this threshold approach may not be of concern, since the principle of ion exchange is that a concentration equilibrium must exist between compartments, and concentrations in the water compartment must be extremely high to have 100 % exchange sites occupied. The exchange partner would then also become high in concentration in the water. For precipitation this also does not hold either, since precipitation may continue forming many layers on the surface of PM. For binding, the threshold model may hold best, but binding also has exchange reactions that, in theory, maintain an equilibrium distribution of chemical species between compartments.

3.2.3.1 Results of K_ds with sediment and radionuclide properties

Between 1965 and 1973 intensive studies were carried out to understand the major reactions of fallout radionuclides with various ocean sediments. The studies comprised the cooperative efforts of 20 national laboratories together with the IAEA marine laboratory in Monaco. Radionuclide experiments were carried out on 35 out of 60 characteristic ocean and coastal sediments (Duursma and Bosch 1970; Duursma and Eisma 1973). Sediment composition analyses were mainly carried out at the Netherlands Institute for Sea Research (NIOZ, Texel, Nl) and specific element analyses were done in the Battelle Laboratories, USA.

As explained in section 2.2.2.2, various sediment properties such as grain size, specific surface (Sp.S.), base-exchange capacity (BEC), etc. were determined, where only a certain proportionality could be detected between K_ds and BEC or Sp.S. (Fig. 3.11).

For ^{137}Cs, when correlated to BEC, potassium (K) and illite concentrations in sediment, the correlation coefficients (R) were 0.3 for BEC and only 0.05 and 0.01 for illite and potassium, respectively (Fig. 3.12).

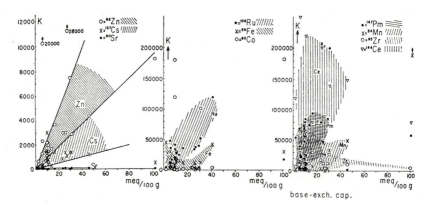

Fig. 3.11. Relationship of K_ds (ml/ml), as determined for ten radionuclides and 35 ocean sediments with the thin-layer technique, and the Base-Exchange Capacity (BEC).

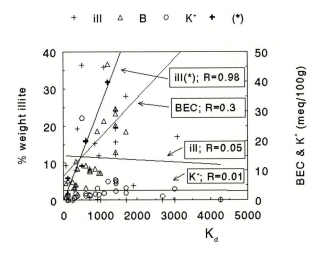

Fig. 3.12. Correlation of [137]Cs K_ds with BEC, illite content and potassium content (K) for 30-33 coastal and ocean sediments. The correlation marked with (*) considers the sediments chosen for the experiments given in Fig. 3.9 and 3.10.

In contrast four sediments, selected by Aston and Duursma (1973), indicated that K_d (^{137}Cs) correlated with the % of illite with a R = 0.98. This demonstrates that statistical correlations may depend on preselection of data and their number used.

3.2.3.2 Other correlations

K_ds - clay minerals

L. van Geldermalsen, Centre of Marine and Estuarine Research (CEMO), Yerseke (Nl) conducted a statistical treatment of correlations for the same 35 ocean sediments and K_ds of 9 radionuclides. The results are given in Table 3.1.

[90]Sr-K_d shows no correlation with any of the sediments. This must be due to complete reversibility of Sr exchange, given by the 100% leacheability and K_d-^{90}Sr/K_d-stable Sr = 100% (see Table 2.2). [137]Cs has indeed a significant (but 2nd degree) correlation with illite, but this is also the case with mica, montmorillonite, kaolinite, and even quartz and calcite. Thus, the matter is more complicated than a simple correlation with only illite. [65]Zn-K_d has only a 2nd degree correlation with illite, and additionally only 2nd degree correlations with mica, montmorillonite and quartz.

Table 3.1. Significance of correlation between K_ds of nine radionuclides with the specific mineral surfaces of 35 coastal and ocean sediments. The significance is calculated on those K_ds from which the most extreme has been deleted.

Nucl.		Deg.	Ill.	Mic.	Mon.	Chl.	Kao.	Qua.	Cal.	Ara.	Dol.	Fel.
^{90}Sr	1^e	-	-	-	-	-	-	-	-	-	-	
	2^e	-	-	-	-	-	-	-	-	-	-	
^{137}Cs	1^e	-	-	-	-	**	**	*	-	-	-	
	2^e	**	**	**	-	*	-	**	-	-	-	*
^{106}Ru	1^e	-	-	-	-	*	-	-	-			
	2^e	-	-	-	-	-	-	-	-			
^{59}Fe	1^e	-	-	-	-	-	-	-	**	-	-	
	2^e	-	-	-	-	-	-	-	***	-	-	
^{65}Zn	1^e	-	-	-	-	-	-	*	-			
	2^e	**	*	*	-	-	**	*	-			
^{60}Co	1^e	-	-	-	-	-	**	**	-			
	2^e	-	**	*	-	-	-	**	-			
^{147}Pm	1^e	-	-	-	-	*	**	-	-			
	2^e	-	-	-	-	*	-	-	-			
^{54}Mn	1^e	-	-	-	***	*	-	-	-			
	2^e	-	-	-	***	-	-	-	-			
^{95}Zr/ Nb	1^e	-	-	-	-	-	-	-				
	2^e	-	-	-	-	-	-	-				

*: $p < 0.05$; **: $p < 0.01$; ***: $p < 0.001$. Not significant is -: $p > 0.05$ (given in Duursma et al. (1983), reproduced also by Duursma and Bewers (1986). Nucl. = nuclide, Deg. = degree, Ill. = illite, Mic. = mica, Mon. = montmorillonite, Chl. = chlorite, Kao. = kaolinite, Qua. = quartz, Ara. = aragonite, Dol. = dolomite, Fel. = feldspar.

The results of statistical analysis of this large data set suggest that although some correlations may be found, there exists no consistent pattern of correlation between mineral type and K_ds. As a result, extrapolation of these correlations to other sediments would be speculatory.

3.2.3.3 K_ds and time of sorption

3.2.3.3.1 One year sorption and leaching

Most K_d determinations are carried out on time scales from 4-7 hours (Balistrieri and Murray 1983; Jannasch et al. 1988; Sioud, 1994) to three weeks (Duursma and Bosch, 1970). Both determine only those partitions which occur in this time span, although apparent equilibria may be attained, depending on the kind of element and particulate matter involved. For adsorption in a time-span of a few hours, the kinetics (Jannasch et al. 1988) show a very rapid uptake of radionuclides in the first minutes with high rate constants, such as for ^{113}Sn, ^{230}Th and ^{46}Sc, this to be less rapid for ^{65}Zn with low rate constants (Table 3.2).

Table 3.2. Rate constants determined for a two sorption reaction model (Nyffeler et al. 1984).

Rate constants	Sn	Th	Sc	Zn
k_1 (d^{-1})	30	5.0	2.0	0.07
k_2 (d^{-1})	-0.5	0.005	-0.005	0.16

A major reason for a lack of information on K_ds, determined for periods of months or years is the problem of setting up the proper experiments which would provide reliable results.

The experiment of Ros Vicent et al. (1974) to study in detail ^{65}Zn uptake by two kinds of sediment revealed that ^{65}Zn accumulated for a very long period of 200 days (Fig. 3.13), whereas the amount of adsorbed ^{65}Zn remained constant from the start. This adsorbed fraction was thought to be the fraction which was easily leachable, while the additionally accumulated fraction was considered as a residual one, representing inclusion of ^{65}Zn in the crystal lattices of the sedimentary particles. Whether this observed phenomenon can be interpreted generally, even for ^{65}Zn, is a matter of consideration. As we will see below (section 3.2.3.2.2). K_ds, as determined during three week sorption, may well be applicable to calculated one year diffusion.

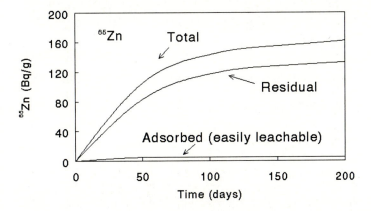

Fig. 3.13. ^{65}Zn activity of different leacheable fractions in counts per min/g dry clay sediment, as function of time of a 9 months *in situ* experiment (Duursma et al. 1975).

Here again, we face the fact, as earlier stated, that K_ds are empirical factors which are difficult to extrapolate from one situation to another, but nevertheless represent the partition of elements for the situation determined.

3.2.3.3.2 One year diffusion and sorption

The same result may not, however, be found under all conditions. As will be explained more in detail in chapter 4 on diffusion in consolidated sediments, the diffusion coefficient can be calculated from the K_d by the formula:

$$D = \frac{D_{chloride}}{1 + K_d} \qquad\qquad (3.12)$$

A diffusion experiment, lasting one year showed a certain correspondence between the determined diffusion coefficients and those calculated on the basis equation (3.12), where K_ds were determined with the thin-layer technique over three weeks (Table 3.3).

Table 3.3. Calculated (by equation 3.12) and determined (instant source technique, see 4.2.2.) diffusion coefficients after a diffusion experiment lasting one year.

Nuclide	Diffusion coefficients ($\times 10^{-10}$ cm^{2}s^{-1})			
	Calculated range		Determined range	
^{90}Sr	5000	30000	7000	11000
^{90}Y		60		
^{137}Cs	14	30	80	110
^{106}Ru	4	5	15	150
^{59}Fe	1.2	6	20	130
^{65}Zn	2.4	24		10
^{60}Co	1	>50	10	20
^{147}Pm	1.6	2.4	4	10
^{54}Mn	2	6	3	5
^{95}Zr/Nb	0.5	5	6	6
^{144}Ce	1	3	5	20

The results demonstrate that even for ^{65}Zn, the order of magnitudes of the calculated and determined diffusion coefficients are almost the same. It may thus be concluded that for the K_ds, determined in the 3 weeks experiment, remain valid for determining diffusion over longer time scales (up to 1 year).

3.2.3.4 Comparison of radiotracer and stable element K_ds

The comparison of K_ds determined for short period sorption and those of natural constituents in the sea, is a critical domain. Li et al. (1984a) found some agreement, demonstrating that there is a difference of two orders of magnitude between the 'exchangeable element'-K_d and the 'consolidated element' K_d (Fig. 3.14A).

For eight of the radiotracers used, Duursma (1973) found identical differences in order of magnitude (Fig. 3.14B) between the tracer partition coefficients of and those of their stable counterparts.

The difference between radionuclide K_ds and stable metal K_ds of ocean sediments is in fact due to surface-available elements which exchange with radionuclides and elements locked more tightly into the structure of the matrix and are therefore not readily available for exchange (see also section 3.2.1.1.5).

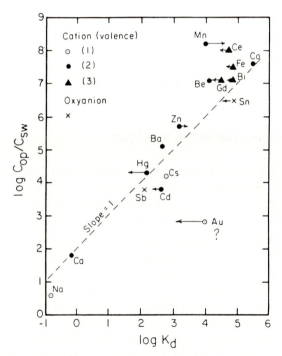

Fig. 3.14A. Comparison of the natural partition coefficient of some elements with 'adsorption' K_ds, determined for various radiotracers and a red clay suspension between 400 and 1700 mg/l (Li et al. 1984a).

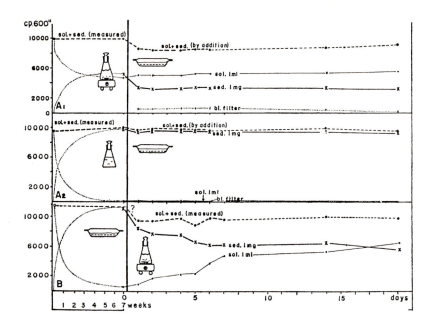

Fig. 3.16. Experiment to compare the thin-layer and suspension technique by interchanging.

The main cause of the differences probably lie in the changing sediment properties due to grinding during magnetic stirring (A1). In order to obtain reliable results with the suspension technique such stirring should be avoided; it is sufficient to shake temporarily by hand.

3.2.4.2 Dialysis/suspension techniques

The results of the determination of K_ds with particulate matter and phytoplankton have been compared for a suspension of Mediterranean sediment (collected near Monaco) and a phytoplankton culture of *Phaeodactylum tricornutum*.

The $K_{d(PM)}$s and $K_{d(pl)}$s of the nuclides [54]Mn, ^{60}Co, [65]Zn, [109]Cd and ^{137}Cs were determined after only 7 days of contact. As shown in Fig. 3.17. there is a certain degree of agreement in the results.

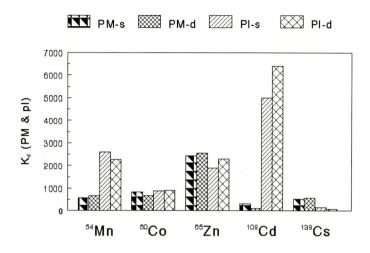

Fig. 3.17. Comparison of K_ds for PM and phytoplankton by filtering a suspension (PM-s and Pl-s) after 7 days contact and the dialysis technique after equilibrium (PM-d and Pl-d) is attained.

3.2.4.3 K_ds of plutonium

The determination of distribution coefficients of plutonium isotopes is not an easy task. For the isotopes ^{238}Pu and $^{239/240}$Pu the handicap is their α radiation which makes measurements very expensive and time consuming. On the other hand experiments with the γ-radiating isotope ^{237}Pu are easy to carry out, but the isotope is short-lived (T½ = 45 days) and is very expensive to produce by cyclotron.

Due to a cooperation between the Fishery Radiobiological Laboratory, UK, of Lowestoft and the IAEA Laboratory (Pentreath 1978) an amount of ^{237}Pu was obtained for a number of sorption experiments with Mediterranean sediment, and the ^{237}Pu was separated into three valence states III, IV and VI by Dr. A. Murray (IAEA Laboratory, Monaco 1977). Furthermore two sorption techniques were applied and sorption was carried out under oxic and anoxic conditions. The results are presented in Table 3.4.

These values have only indicative value for sorption as observed in nature, and some higher values might be expected for sorption occurring over longer time scales.

Table 3.4. Plutonium-237 K_ds with Mediterranean sediment as determined for different valence states of ^{237}Pu, two sorption techniques and under oxic and anoxic conditions. The K_d ($x10^4$) is given in ml/g; the determination lasted three weeks and countings were corrected for decay (Duursma and Parsi 1976).

	Oxic conditions pH: 7.8-8.0		Anoxic conditions pH: 7.8-8.0	
Techniques→ Valence↓	Settling	Thin-layer	Settling	Thin-layer
III	1.6	2.1	1.9	
IV	1.8	1.9; 1.5	1.3	9.4
VI	1.3	5.7	2.2	

Recent measurements by Molero et al. (1995) demonstrated that K_ds, determined for the same dimensions (l/kg) as Duursma and Parsi (1976), were higher and ranged from 8-23x10^4 (Table 3.5), values which are of the same order as given by Sholkovitz et al. (1983) and Sholkovitz and Mann (1984) for Buzzards Bay and by Buesseler and Sholkovitz (1987a,b) for shelf, slope and deep-sea sediments of Atlantic (Table 3.6), taking into account that K_ds (l/kg = ml/g) are equal to K_ds (g/g).

Table 3.5. K_ds (x 10^4; l/kg) as determined for the distribution of ^{239,240}Pu and ^{241}Am between dissolved and particulate nuclide in surface sea water samples collected along the Spanish Mediterranean coast (Molero et al. 1995).

Location	^{239,240}Pu		^{241}Am	
	Mean	Standard Deviation	Mean	Standard Deviation
Cabo de Creus	11	6	30	10
Tossa	23	4	40	10
Barcelona	8	3	120	20
Vandellós I	10	4	70	20
Vandellós II	12	8	100	30
Delta del Ebro	13	8	30	10
Denia	14	2	170	40
Cabo de Palos	21	5	130	60
All Mean (n=8)	14	5	90	50

Table 3.6. K_ds (g/g) as determined for the distribution of ^{239,240}Pu between pore water and sediment particles, as determined by Sholkovitz et al. (1983), Sholkovitz and Man (1984) and Buesseler and Sholkovitz (1987a&b).

Location	K_d ($\times 10^4$)
Buzzards Bay	3-12
Shelf/slope/deep-sea of Atlantic	2-23

4 Diffusion principles

4.1 Theories

Ever since contamination of the marine environment began, the processes by which the concentrations of these contaminants increase or decrease have been a subject of investigation. Currents, with their turbulent (eddy) diffusion, are the main cause of distribution and dilution in marine systems, while sorption, precipitation and co-precipitation processes may cause local accumulations, usually in bottom sediments. A process of molecular diffusion cannot be of great significance in the sea, but remains crucial for migration of molecules or ions in interstitial water of bottom sediments, where movement of the pore water itself is negligible.

The kinetics of this process of migration in natural sediments is presented as a diffusion process, founded on the basic diffusion law for finding experimentally the major factors which characterize the transfer of contaminants in bottom sediments.

In principle, sediments of lakes, rivers and seas are saturated with water, and when there is no horizontal pressure gradient or bioperturbation caused by meiofauna living in the pore water and larger zoobenthos, causing reworking of the sediment, there is no mechanism of water transport inside the sediment. Then the concentrations in the pore water are determined by the process of migration, and the variation with time and distance can be calculated with help of the diffusion laws.

It is fairly easy to set up a mathematical description of a certain situation in which migration, adsorption or absorption is taking place, but this usually leads to equations which are not easily solved. It is more difficult to present the situation in such a way that a maximum simplicity of mathematical computation coincides with a maximum simplicity of experimental design and field conditions. A practical example of this can be given. Let us suppose that radioactive waste is put in a bore hole and that it can diffuse horizontally between two impermeable layers into a water-saturated sandy sediment. In order to find out the diffusion coefficient, which is the principle factor of diffusion, the theory should be able to indicate the simplest way of doing this experimentally. As is explained later, this can be done for example by putting at one instant a quantity (not necessarily known) of radiotracer in the bore hole and determining as a function of time the radioactivity in the same sediment layer in another bore hole some distance away.

Simple laboratory and *in situ* techniques are developed to determine the diffusion coefficients in pore water of sediments irrespective of sorption of diffusing substances to sedimentary particles. Partial changes of the molecular diffusion coefficient of migrating molecules or ions in pore water may be influenced by the encountered obstacles by which the mean free path is longer, temperature, charge of the solid-liquid interfaces of sediment particles and changes in properties of the interstitial water such as surface tension and viscosity. In order to determine their impact, a number of simple laboratory experiments were set up to determine these changes as function of grain size of sediment particles for the mean free path, the electrical potential retention in pore water, surface action on van der Waals or Coulomb forces, interfacial tension, viscosity and zeta potential. Since temperature, pressure and salinity are parameters which vary in ocean bottom sediments, their impact on diffusion was equally estimated.

4.1.1 Diffusion from constant and instantaneous sources

Diffusion of substances, either in a liquid (fully at rest) or in a solid are given by Fick's first equation:

$$F = -\phi D \frac{\delta C}{\delta x} \tag{4.1}$$

where F is the flux of material (mol.s^{-1}) through 1 cm^2, ϕ the porosity (Ullman and Aller (1982), D is the diffusion coefficient (cm^{2}s^{-1}), C the concentration (mol.cm^{-3}), and x the path (perpendicular on the 1 cm^2 surface) of the diffusion in cm.

Fick's second equation follows from (4.1):

$$\frac{\delta C}{\delta t} = D \frac{\delta^2 C}{\delta x^2} \tag{4.2}$$

These equations, transformed in applicable formula's (Duursma and Hoede 1967), are used to determine apparent diffusion coefficients for compounds migrating into or out of marine sediments. Two examples are:

$$C(x,t) = C_0 \, erfc \frac{x}{2\sqrt{DT}} \tag{4.3}$$

and

$$C(x,t) = \frac{s}{\sqrt{(4\pi Dt)}}\, e^{\left(-\frac{x^2}{4Dt}\right)} \tag{4.4}$$

where, for linear diffusion, C(x,t) is the concentration in the sediment at a certain depth x and time t from the start of the diffusion, with C_o as a constant source concentration for equation (4.3) and s is the amount of instantaneous source for equation (4.4). Erfc is the symbol for the complementary error function, and D is the diffusion coefficient. D can be determined by placing data from simple experiments into the framework of Figs. 4.1 and 4.2.

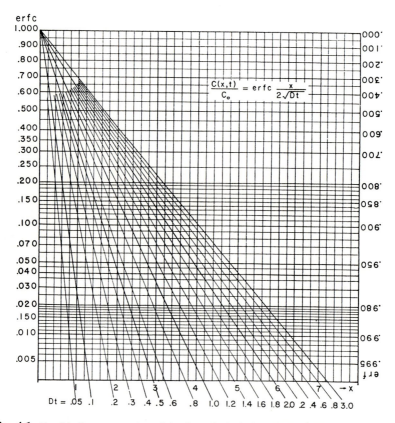

Fig. 4.1. Graphical representation of the formula (4.3), for different values of Dt. By plotting the results as $C(x,t)/C_o$ against x, Dt can be estimated by interpolation. When t is known, D can be calculated. The left Y-axis (erfc = complementary error function) is the reverse of the right Y-axis (erf = error function).

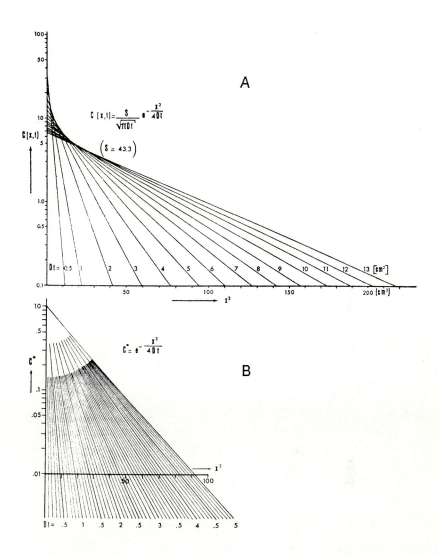

Fig. 4.2. A: Graphical representation of formula (4.4), where C(x,t) is plotted against x^2. This will result in straight lines for different values of Dt. The lower graph B, as transparent photocopy, should be superimposed over an experimental plot with identical axis to determine Dt.

4.1.2 Diffusion and sorption

Using equation (2.11) as a starting point for the exchange of chemicals, dissolved in pore waters and sorbed onto sedimentary particles, diffusion into

$$\frac{\delta C}{\delta t} + \frac{\delta (K_d\, C)}{\delta t} = D\frac{\delta^2 C}{\delta x^2} \tag{4.5}$$

which becomes:

$$\frac{\delta C}{\delta t} = \frac{D}{1+K_d}\frac{\delta^2 C}{\delta x^2} \tag{4.6}$$

where D is the diffusion coefficient of the dissolved chemical in pore water.

This D is a function of both the molecular diffusion coefficient and the matrix of the sediment, i.e. the porosity ϕ (Ullmann and Aller 1982). For simplicity, D will be used here as the diffusion coefficient of the chloride ion, which can easily be determined with the help of its radioactive tracer (^{36}Cl) in the form of KCl (as we will see in section 4.3.3.1 it is essential for determining D_{Cl} that the cationic and anionic mobilities are equal such as for K^+ and Cl^-).

4.2 Simple methods of determining diffusion coefficients

4.2.1 Constant source technique

Plastic tubes, filled with sand and closed with a fine plankton net (size 64 μm) are hung in a large bottle, filled with water spiked with a tracer. The volume must be large enough to guarantee a constant concentration of the tracer for the duration of the experiment. In case of a density difference between the solution and the pore water, the tubes should be hung in such a way that convection is avoided (Fig. 4.3).

After exposure, the concentration of the migrated tracer is determined by freezing, lyophilization, cutting and radio activity counting (Fig. 4.4). The results are plotted according to Fig. 4.1.

4.2.2 Instantaneous source technique

This technique has the advantage of being easier to carry out than the constant source technique, since uptake is independent of the amount of added instantaneous source. Further the experiment does not require maintenance of the source. A small amount of tracer is added to a tube filled with sediment and closed with a stopper to avoid evaporation of the supernatant and pore water (see Fig. 4.4). The results are plotted according to Fig. 4.2.

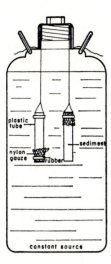

Fig. 4.3. Technique for determining the diffusion coefficient in a sediment with a constant source.

4.3 Experiments on hindrance of diffusing ions in sediments

Diffusion processes, which occur in pore (or interstitial) water of sea bottoms, are retarded by matrix hindrance and reactions of dissolved species with the sedimentary particles. Alternatively, diffusion will be enhanced by any movement of pore water, caused by either meiofauna living in these pore waters or by macrozoobenthos causing bioperturbation. For the following experiments these bioperturbations are *not* taken into account.

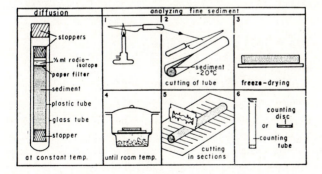

Fig. 4.4. Technique for determining the diffusion coefficient in a sediment with an instantaneous source and subsequent processing of sediment subsamples.

The factors having a possible impact on retardation are (i) mean free path and porosity, (ii) temperature, (iii) electrical potential retention (concentration and dissociation effects), (iv) surface action and connected interfacial tension, viscosity and zeta Potential (zeta potentials were measured by V.Pravdic, Zagreb, Croatia).

4.3.1 Mean Free Path

Dissolved chemical species in motion must move around sedimentary particles, and thus their distance travelled in direction x is longer. For close packed uniform particles, the free path is about 1.4x longer, independent of grain size (Fig. 4.5), resulting in a diffusion coefficient of $(1.4)^2 = 2$ lower.   In reality the mean free path may or may not be dependent of the grain size for sieved sandy sediment fractions, because real packing is less than for idealized spheres.

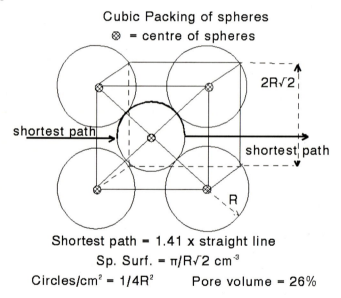

Shortest path = 1.41 x straight line

Sp. Surf. = $\pi/R\sqrt{2}$ cm^{-3}

Circles/cm^2 = $1/4R^2$ Pore volume = 26%

Fig. 4.5. Mean free path in cubic packed identical spheres. R = radius of spheres. The shortest path per cubic unit is $2R(\sqrt{2}-1)+\pi R = 3.97R$, indicating that the free path is $(3.97R/2R\sqrt{2}) = 1.40$ x straight length. The surface of the spheres $(16\pi R^2)$ per cubic unit $(16R^3\sqrt{2})$ is $\pi/R\sqrt{2}$ (proportional to R^{-1}). The interspace volume of cubic unit $(16R^3\sqrt{2})$ minus that of the parts of the (4) spheres within one cubic unit $((16/3)\pi R^3)$, which gives a pore volume of 26%. The effective solid cross-section surface is the surface of one lateral face of the cubic unit $(8R^2)$ minus that of 2 circles intersected by a lateral face $(2\pi R^2)$ divided by 8R^2, resulting in 21.5%.

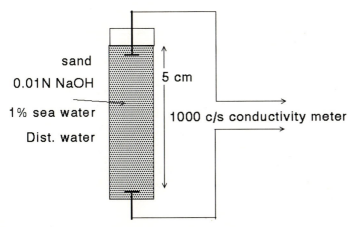

Fig. 4.6. Technique to determine the relative specific conductivity of pore water of sieved sand fractions.

An average, but consistent pore volume of 41 and 38% was determined for various fractions of Mediterranean and North Sea sand, respectively, which is higher than the 26 % pore volume, calculated for cubic close packing and 32.8 % of hexagonal close packing (Duursma and Bosch 1970). For the same North Sea sieved sand fractions it was demonstrated that the electrical conductivity, as measured according to the scheme given in Fig. 4.6., produced a constancy of $1.16 \pm 0.1 \times 10^4$ Ohm (Fig. 4.7), which confirms the independence of porosity with grain size.

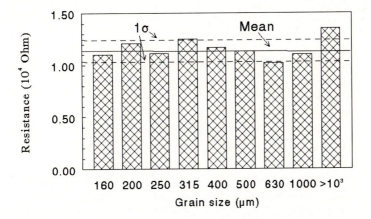

Fig. 4.7. Measured electrical resistance of various grain-size fractions of sand, loaded with 0.01 N NaOH, 1% sea water in distilled water, according to the technique given in Fig. 4.6.

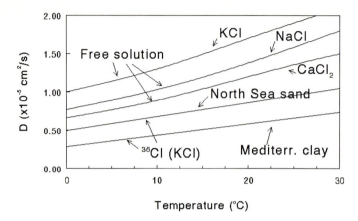

Fig. 4.8. Temperature dependence of the diffusion coefficients of KCL, NaCl, CaCl$_2$ in free solution (Lide 1993) and that of ^{36}Cl in Mediterranean pelagic clay and North Sea sand.

4.3.2 Temperature

Free movement of dissolved molecules is enhanced at higher temperatures (T), causing higher diffusion coefficients (D) in free solution. The average dD/dT is 0.3x10^{-5} cm^{2}s^{-1}/°C (Fig 4.8).

Diffusion coefficients in ocean sediments, where the average temperature is about 2 °C, will therefore differ by 3x10^{-5} cm^{2}s^{-1} from coastal sediments with average temperatures of about 12 °C.

4.3.3 Electrical potential retention in pore water

4.3.3.1 Charge of ions

Diffusion of charged ions is accompanied by the diffusion of their oppositely charged counterparts. For example Cl$^-$ diffusion induces either Na$^+$ or another cation, since a large electrical potential will otherwise be built up. The question regarding the influence of this 'sharing' has either an effect of increase or decrease on the diffusion of Cl$^-$. Indeed, the diffusion coefficients of Cl$^-$ may depend on the counterpart (e.g. Na$^+$, K$^+$, Ca^{++}, Fe^{+++} or H$^+$; Fig. 4.9 and 4.10). Fe^{+++} and Ca^{++} are surrounded by larger H$_2$O clusters than Na$^+$, K$^+$, and H$^+$. This is also indicated by their equivalent conductances.

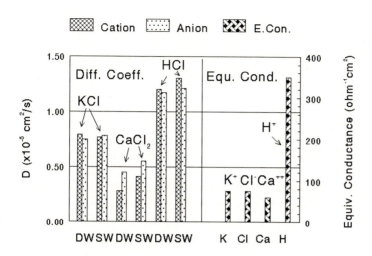

Fig. 4.9. Diffusion coefficients as determined separately of KCl, CaCl₂ in North Sea sand and HCl in Mediterranean pelagic clay. DW = distilled water; SW = sea water.

The result is that H$^+$ (which is only a proton and can move very quickly) gives Cl$^-$ the highest diffusion coefficient and Fe^{+++} the lowest. For K$^+$, which has an equal ionic radius as Cl$^-$, the diffusion coefficients of both anion and cation are almost equal.

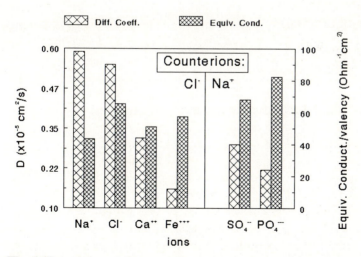

Fig. 4.10. Diffusion coefficients of three cations, with Cl$^-$ as anion and three anions, with Na$^+$ as cation, determined in 125 μm sand. For comparison the equivalent conductances par valency are given.

The conclusion is that this effect must always be taken into account, unless it is negligible with respect to the effect of chemical sorption, given by the K_d of equation (4.6).

4.3.3.2 Concentration of salts

In case diffusion is studied in concentrated salt solutions, part of the salt will become undissociated and therefore uncharged.

As a consequence, undissociated salt will diffuse faster because it has only a low charge as dipole and therefore is less attached to water clusters than its charged individual cations and anions. This phenomenon was confirmed with a diffusion experiment with $SrCl_2$ (Fig. 4.11).

4.3.4 Surface action

4.3.4.1 Specific surface

Van der Waals or Coulomb forces on sediment particle surfaces can influence the rate of migrating dissolved species. So there will be a retention depending on the specific surface area of the sedimentary particles. In such a case, the diffusion coefficient D should be lower for a higher specific surface, proportional to $1/d$, where d is the grain size (see caption of Fig. 4.5). This was indeed found for ^{36}Cl diffusion in different fractions of equal grain size (Fig. 4.12).

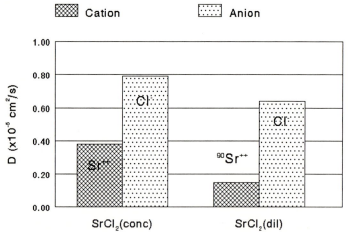

Fig. 4.11. Diffusion coefficients of $SrCl_2$, determined for both Sr^{++} and Cl^- under concentrated and very diluted conditions (sea water level) in North Sea sand, with sea water as pore water.

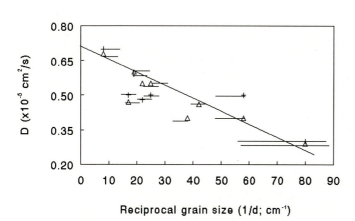

Fig. 4.12. Diffusion coefficients of Cl⁻ ($K^{36}Cl$) as determined for different fractions of sieved sand, plotted against the reciprocal of the grain size. D was determined using two techniques, one with a constant source and one with an instantaneous source.

4.3.4.2 Interfacial tension

The interfacial tension or capillary force, was determined for a similar range of grain sizes as mentioned in Fig. 4.12, by idealizing sand as a system of narrow tubes which dimensions equal to those of pores in the sand, and therefore proportional to the diameter of the grains. The larger the grains of homogenous fractions, the larger the pore sizes. In this simulation we can apply the formula for narrow tubes (Glasstone, 1960) of:

$$h = \frac{2\gamma\cos\theta}{g\rho} \cdot \frac{1}{r} \tag{4.7}$$

where h is the capillary rise in cm, g the acceleration due to gravity, ρ the specific weight of the liquid, r the radius of the tubes, θ the solid-liquid angle of contact and γ the surface tension solid-liquid (interface tension).

Assuming that $(\cos\theta)/(g\rho)$ is the same for all grain-size fractions, and r is proportional to the diameter of the sand particles (d) and A is a constant, the formula becomes:

$$h = \frac{A\gamma}{d} \tag{4.8}$$

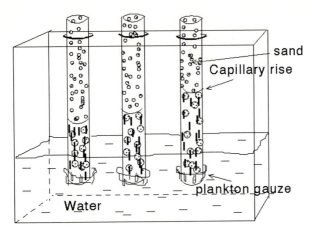

From coarse ==> finer grain sizes

Fig. 4.13. Technique used to determine the capillary rise of water in sieved North Sea sand fractions.

An experiment, carried out as shown in Fig. 4.13, indicated a linear correlation between the capillary rise (h) and the reciprocal of d (Fig. 4.14). In that case it can be concluded from formula (4.8) that the interfacial tension γ is constant, and thus independent from the grain size.

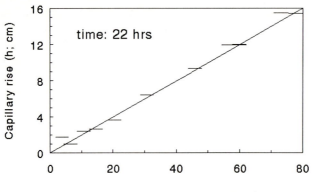

Fig. 4.14. Capillary rise after 22 hours plotted as function of the reciprocal of the grain size d (experiment Fig. 4.13).

4.3.4.3 Viscosity

Since the viscosity of pore water is affected by either solid-liquid interfacial or particle surface tension due to changing grain sizes, diffusion of chemicals in pore water will certainly be affected too. An experiment was set up to investigate the dependency of the pore water viscosity on different grain sizes. Similarly as for the capillary experiment, the sediment was idealized as a system of narrow tubes through which water passes in laminar flow. For these narrow tubes the coefficient of viscosity is given by the Poiseuilles equation (Glasstone, 1960):

$$\eta = \frac{\pi P r^4 t}{8vl} \tag{4.9}$$

where η is the viscosity coefficient, P the driving pressure imparted to the flow through the capillary tube, r the radius of the capillary tubes, t the time of the experiment, v the volume passing in time t and l the length of the tube.

The number of capillary pores in the experimental plastic tube will be $Q/4R^2$, if Q is the surface of the section of the plastic tube and R the radius of the spheres (see also caption of Fig. 4.5). The volume passing through the plastic tube should be multiplied with this factor, with the result that equation (4.9) can be simplified to the correlation known as Darcy's law (4.10):

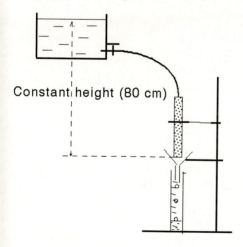

Constant height (80 cm)

Fig. 4.15. Technique used to determine the viscosity of pore water for different grain-size fractions.

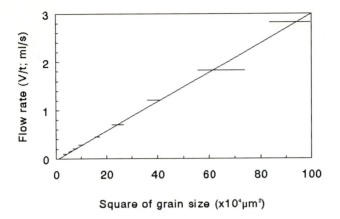

Fig. 4.16. Flow rates in ml/s as determined with the experiment, shown in Fig. 4.15, as a function of the square of the grain size (d^2).

$$\frac{v}{t} = \frac{Ad^2}{\eta} \tag{4.10}$$

where A is a constant and d the grain-size diameter. Plotting the results of the experiment shown in Fig. 4.15, there is indeed a linear correlation (Fig. 4.16) between the flow (v/t) and d^2 until about 1000 μm ($d^2 = 10^6 \ \mu m^2$) grain-size. This implies a constant viscosity coefficient for the interstitial water of these sand size fractions.

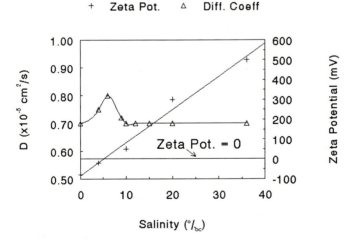

Fig. 4.17. Dependence of diffusion of ^{36}Cl (KCl) in interstitial water of sand on salinity, due to change in the zeta-potential.

4.3.4.4 Zeta Potential

The Zeta Potential or electro-kinetic potential is a measure of the charge of sedimentary particles and is dependent upon the ionic strength of the medium in which the particles are retained. For the salinity range of 0 to 36 ‰ the zeta potential ranges from - 80 mV to +530 mV (Fig. 4.17).

The effect on diffusion is a retention at low or high zeta potentials (irrespective of the sign), whereas diffusion is faster at zeta potentials close to zero. This means that a diffusion coefficient is only slightly higher in low salinity waters.

4.3.5 Effect of hydrostatic pressure on diffusion in sediments

Although water is virtually incompressible (0.458 % at 1000 decibar), which occurs at about 968 m (Cox, 1965), the large hydrostatic pressure in the deep ocean and in the sea bed may affect certain chemical reactions.

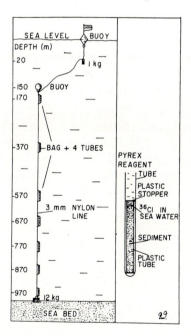

Fig. 4.18. Diffusion experiment of ^{36}Cl (NaCl) in interstitial water of marine sediment under different pressures.

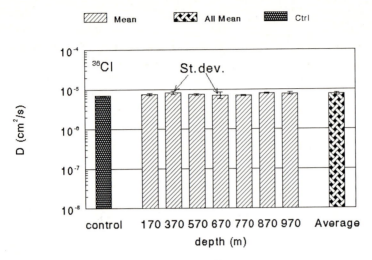

Fig. 4.19. Apparent diffusion coefficients for ^{36}Cl measured *in situ* at various depths. n = 4 per depth.

The prediction of apparent diffusion coefficients for material migrating in deep-sea sediments has most frequently been carried out on the basis of the determination of concentration gradients in deep-sea cores, or by carrying out diffusion experiments in the laboratory under ambient conditions.

In order to measure the change in diffusion coefficients for chloride in marine sediment, Duursma (1977) carried out an *in situ* experiment down to a depth of about 1000 m (Fig. 4.18). The experiment was carried out according to the instantaneous source model (equation 4.4) and lasted seven days. A control at depth zero was carried out in the laboratory.

The results are shown in Fig. 4.19, demonstrating that, within the accuracy of the measurements, no pressure effect could be detected for the diffusion coefficients determined. Extrapolation of these results to greater depths is not possible, but any effects are likely to be negligible. This is also the opinion of Shiskina et al. (1969) in their monograph on halogen diffusion in deep-sea sediments.

4.4. Summary on effects on molecular diffusion in bottom sediments

From the various experiments carried out in relation to molecular diffusion in water saturated sediments of different grain size, it becomes clear that the impact of each of the physical factors on the molecular diffusion coefficient, such as matrix hindrance, temperature, ionic charge, concentration, surface action through van der Waals and Coulomb forces, interfacial tension (capillarity), viscosity, zeta potential and pressure is much less than diffusion plus chemical sorption. Nevertheless, as presented in Table 4.1, some effects are clearly present, others are totally absent. This demonstrates that studies on molecular diffusion in sediments in the absence of sorption must still deal with the physical factors. This is valid for diffusion of such ions or molecules like nitrate, ammonia and oxygen, which do not have chemical sorption to sediment particles.

Table 4.1. Summary of effects on molecular diffusion in water saturated sediments.

Parameter	Effect on Diffusion coefficient	Dependence of grain size
porosity	40-45%	no
hindrance (mean free path)	x $(1/1.41)^2$ = ½	no
temperature	0.3×10^{-5} cm^{2}s^{-1}/°C	no
ionic charge	x 2.8 (Cl$^-$ in HCl/CaCl$_2$)	
concentration	x 1.3 (Cl$^-$ bound in SrCl$_2$ molecule/Sr^{++} & Cl$^-$ ions)	no
vd Waals and Coulomb forces	x 2.3 (100=>1000 μm)	yes
interfacial tension (capillarity)	no: int. tens coef. γ = const.	no
viscosity	no: viscosity coef. η = const.	no

Parameter	Effect on Diffusion coefficient	Dependence of salinity
zeta potential	x 1.3 at 0 zeta pot.	yes
pressure	none for 1000 deciBar (968 m depth)	no

The main conclusions of the experiments presented in this chapter are summarized as follows:

Porosity
- Porosity, when calculated for cubic sphere packing, causes an increase of mean pathway in water of 1.40, which affects the apparent molecular diffusion coefficient, expressed in cm^2/s, with a factor $(1.40)^2$ lower than in free solution,

which means roughly a factor 2. For natural sands, having constant porosities ranging from 38 to 41% for different grain size fractions this factor will be slightly less.

Temperature
- Temperature affects diffusion of ions in a similar proportion as their equivalent conductance, which results in a $\Delta D/°C$ of 0.3×10^{-5} $cm^2 s^{-1}/°C$.

Ionic Charge
- The ionic charge of ions and dipole charge of molecules cause proportional clustering with H_2O molecules, which retard their movement. Diffusion of individual ions are mutually dependent of the diffusion of their counter ion. The higher the valency, the lower is the diffusion coefficient, caused by the above mentioned water clusters. On the other hand the hydrogen ion H^+, which is a single proton of high mobility enhances the diffusion of the counter ion Cl^- with respect to Ca^{++} by a factor of 2.8 for its diffusion coefficient. Molecules of undissociated salts (under saturated conditions) diffuse faster than their independent ions (under very dilute conditions). A maximum difference in diffusion coefficients is of the order of a factor 1.3.

Specific surface
- The specific surface of sediment particles, which is larger for smaller grain size, affects charge forces between the sediment particles. Diffusion coefficients may demonstrate a correlation with the reciprocal of the grain size which is proportional to the specific surface. For sand fractions, ranging from 100 to 1000 μm, this may affect the diffusion coefficients with a factor 2.3, the lowest diffusion coefficient measured for the finest fractions. On the other hand the specific surface does not affect the interfacial tension coefficient and the viscosity coefficient. Thus these physical properties of pore water for coarse and fine sediment are the same. This will therefore not have any effect on the diffusion coefficient.

Charge
- The charge of aquatic sediment particles depends on the ionic strength of the pore water, which may range from -100 mV to 500 mV, when measured as zeta potential (or electrokinetic potential). Freshwater sediment particles are charged negatively and marine sediment particles positively. In the brackish region of 6 ‰ salinity, the zeta potential is zero. Thus the apparent molecular diffusion coefficients are identical in fresh water and marine water above 10 ‰ salinity, but may peak with a factor 1.3x higher at 6 ‰ salinity.

Pressure
- Since water is nearly incompressible, a negligible effect on molecular diffusion is expected. An experiment confirmed this for $^{36}Cl^-$ diffusion until 1000 deciBar (968 m). At this depth the compressibility is only 0.458 %, which means for deep ocean sediments of 5000 m the compression is 2.4%.

Remark: Most of these effects will fall within the noise of the determination of diffusion coefficients when chemical reactions (sorption-desorption) take place between the migrating molecules or ions and the sediment particles.

5 Organochlorines

5.1 Introduction

Organochlorine contaminants are those environmentally persistent organics which require severe monitoring control and countermeasures to avoid world-wide hazardous effects. The pesticide DDT (Cl-Phenyl-HCCCl$_3$-Phenyl-Cl) was first applied in the early 1940s and seemed to have a bright future. The pesticide made complete human populations lice-free, and it was applied to control a great number of pestilent insects, among which the malaria mosquito was the largest target. The pesticide was effective due to its persistence, and seemed to have little effect on man, due to its low oral and dermal toxicity of 200 mg/kg bodyweight and 3000 mg/kg, respectively (Duursma and Marchand, 1974). In comparison, the oral toxicity of the non-persistent organo-phosphorus insecticide Parathion is 0.2 mg/kg.

In 1962, the book 'Silent Spring' by Rachel Carson (Carson 1962) brought a large change in public and scientific opinion on the use of persistent pesticides. DDT and its metabolites were found as residues in lipids of remote populations of land animals and birds, and a coincident acute decrease of predators was observed. Birds of prey (raptures) and seals seemed especially susceptible to DDT contamination. The idea of accumulation through the food chain was launched and has persisted as a major environmental principle.

Since DDT has been banned in many countries from the early 1970s, the DDT concentrations have decreased in most environmental samples and attention is now mainly focused on polychlorinated biphenyls (PCBs), another family of organochlorines, largely produced and applied in the last 40 years as a nonflammable grease or oil. They are used as isolation liquid in electrical transformers and grease in mines among other applications.

PCBs are even more persistent than DDT and have a similar toxicity: in vivo the ED-50 inhibition body weight gain is 750 and 1120 μmol/kg (234 and 350 mg/kg) for the PCB congeners No. 105 and 118, respectively (Goldstein and Safe 1989). Due to toxicity and persistence, a ban on production and use was insisted in the last decennia. There is, at present, a reserve of about 1.2 million ton of PCBs, of which <62% is in stock or in use, 31% in the environment (3/5 of this in the marine environment), while <3.2 % is degraded or burned (Tanabe, 1988). The option of burning PCBs as a possible disposal method has not yet been applied extensively, due to the very high temperatures required, and the possible liberation of toxic byproducts to the atmosphere.

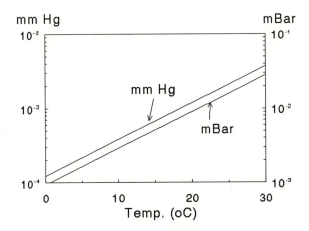

Fig. 5.1. Partial pressure of 4,4'PCB as function of temperature (Lide, 1993).

The potential danger of these PCBs is the fact that even being only slightly volatile, they can slowly evaporate (Doskey and Andren 1981; Burkhard et al. 1985a,b,c; Erickson 1986; Duinker and Bouchertall 1989) into the atmosphere and thus contaminate the complete earth's compartments of water, sediments, atmosphere and biota (Eisenreich and Johnson 1983).

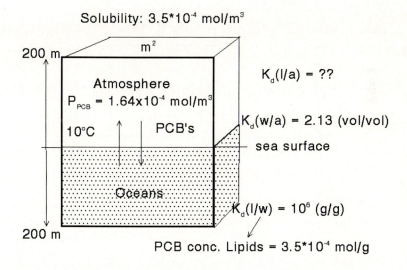

Fig. 5.2. Potential saturation equilibrium of 4,4'PCB between a 200 m layer of atmosphere (of 1 atm.) and a 200 m layer ocean. l/a = lipid/atmosphere, w/a = water/atmosphere and l/w = lipid/water.

The partial pressure of 4,4'PCB (Fig. 5.1.) and its solubility in the oceans (Fig. 5.2) can be used to determine the potential saturation amounts in the atmosphere (Fig. 5.3). Although the world stock of PCBs is composed of higher chlorinated biphenyls than 4,4'PCB, it demonstrates what might happen if the world stocks become open sources.

The distribution of PCBs in air and water compartments can be evaluated on the basis of Mackay (1982)'s data, demonstrating that the solubility of congeners is a function of K_{OW} (equation 5.6 and Fig. 5.5) and the partial pressure of the congener P_{PCB} in air. For example K_{OW} of 2,4,5,2',4',5'-PCB has a value of $7x10^6$, where its solubility in water is $4x10^{-3}$ μmol/l. At this saturation concentration of $4x10^{-3}$ μmol PCB/l (or $4x10^{-6}$ mol/m^3), a vapour pressure can be derived from equation 5.1 (Burkhard et al. 1985c):

$$P = H \cdot C \tag{5.1}$$

where H, the Henry's Law constant ranges for PCB congeners along their number of chlorines. Taking as average for PCBs at 25 °C (Burkhard et al. 1985b), $H = 4.0x10^{-4}$ atm-m^{3}mol^{-1}, P becomes for 2,4,5,2',4',5'-PCB $1.6x10^{-9}$ atm or $1.6x10^{-11}$ Pa. Such a partial pressure represents a congener concentration in air of:

4, 4' PCB; M=223
At 10°C: Pprt=0.00368mBar; Conc.=0.000164 mol/m^3
Max. in atm: $6.65x10^{14}$mol = $1.48x10^{11}$ton

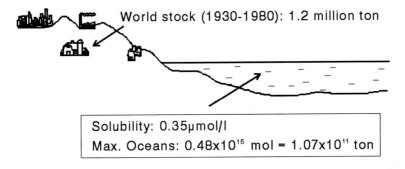

World stock (1930-1980): 1.2 million ton

Solubility: 0.35μmol/l
Max. Oceans: $0.48x10^{15}$ mol = $1.07x10^{11}$ ton

Fig. 5.3. Calculated potential atmospheric and oceanic equilibrium concentrations of 4,4'PCB.

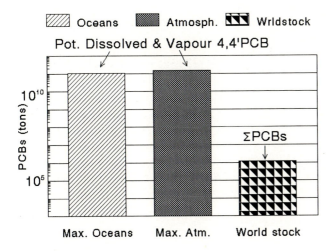

Fig. 5.4. Calculated potential atmospheric and oceanic equilibrium concentrations of 4,4'PCB. Ibid as in Fig. 5.3.

$$C_{(PCB\ in\ air)} = \frac{P_{PCB}}{RT} \qquad (5.2)$$

where R is the gas constant equal to 8.2×10^{-5} atm m³ mol⁻¹ K⁻¹ (K is degrees Kelvin). At 10 °C or T=283 K this gives:

$$C_{(PCB\ in\ air)} = \frac{1.6 \times 10^{-9}}{8.2 \times 10^{-5} \times 283}\ mol \cdot m^{-3} = 6.89 \times 10^{-3}\ mol \cdot m^{-3} \qquad (5.3)$$

From this result the equilibrium distribution factor of PCBs saturated in water and in equilibrium with air can be determined:

$$K_{d(water-air)} = \frac{[PCB\ in\ water]}{[PCB\ in\ air]} = \frac{4 \times 10^{-6} mol \cdot m^{-3}}{6.89 \times 10^{-5} mol \cdot m^{-3}} = 5.8 \qquad (5.4)$$

This value is lower than we will observe later for a variety of ocean and atmosphere sampling stations (*in situ* $K_{d\ (water-air)}$ = 122; Fig. 7.20B). Knowing the approximate K_d of PCB partitioning between lipids of (aquatic) organisms and water of 10^6 ($\mu g.g^{-1}/\mu g.g^{-1}$), which is equal when given in mol.m⁻³/mol.m⁻³, Fig. 5.10) and using the reciprocal value of equation (5.4), the distribution coefficient between PCBs in lipid and air becomes:

$$K_{d(lipid-air)} = \frac{K_{d(lipid-water)}}{K_{d(air-water)}} = \frac{10^6}{0.172} = 5.8x10^5 \ (mol \cdot m^{-3}/mol \cdot m^{-3}) \quad (5.5)$$

or $1.22x10^7$ in case the *in situ* $K_{d \ (water-air)} = 122$ is applied instead.

In the following sections the behaviour of PCBs in the marine environment will be presented (section 5.2) and whether or not the hypothesis on accumulation through the food chain is correct (section 5.3). Attention will be given to a case where pesticides are applied in tropical aquaculture (section 5.4). As a consequence of the high distribution coefficient from atmosphere to lipids, the atmosphere may be suspected as the principal contamination pathway of DDT to terrestrial organisms. This hypothesis will be elaborated on for humans in section 5.5.

5.2 PCB and DDT behaviour in water, PM and sediment compartments

PCB congeners are non-polar compounds and have therefore a low solubility in water, adsorb to solid surfaces (being less-polar than water) and 'dissolve' easily in lipids. Their degree of non-polarity is defined by the partition ratio between water and octanol, giving a partition coefficient (K_{ow}) defined as:

$$K_{OW} = \frac{[PCB]_{octanol}}{[PCB]_{water}} \quad (5.6)$$

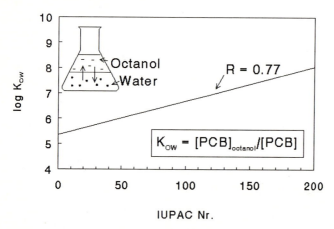

Fig. 5.5. Correlation of PCB IUPAC numbers with K_{ow} (after Brownawell and Farrington, 1985).

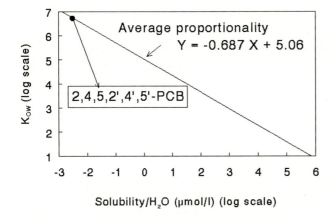

Fig. 5.6A. Average correlation of solubility of various organic compounds in octanol and water (after Mackay 1982).

Each PCB congener has a specific K_{ow} (Rapaport and Eisenreich 1984), from which the log K_{ow} corresponds proportionally (correlation index $r = 0.77$) to its IUPAC Nr. (Brownawell and Farrington 1985; Fig. 5.5).

Similarly K_{ow} is indicative of the solubility of organic compounds in water, demonstrating a correlation when plotted on log scales (Fig. 5.6A). Combining the results from Fig. 5.5 and 5.6A we obtain a correlation between the solubility of PCB congeners in water and their IUPAC numbers (Fig. 5.6B). Although this correlation lacks complete proportionality ($r < 0.77$), it supplies a basis for approximate calculations on the basis of the water solubility of the various PCB congeners.

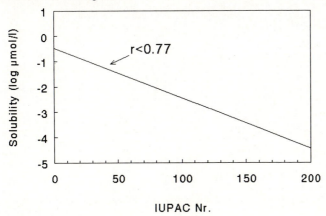

Fig. 5.6B. Correlation between the solubility of PCB congener and their IUPAC Nrs. as obtained from the correlations of Fig. 5.5 and 5.6A.

5.2.1 Partitioning of PCBs between water and PM

The principle of a partition coefficient K_d, as defined by equation (2.10), is used for studying the partitioning of PCB congeners between dissolved and particulate compartments:

$$K_d = \frac{\mu g \ PCB/g \ (dry) \ PM}{\mu g \ PCB/g \ water} \qquad\qquad (5.7)$$

A study was made to determine K_ds in an estuarine system where a large variety of salinities, PM concentrations and PCB concentrations occurred.

As given in Fig. 5.7, log K_d (left bar) is at the level of about 5 (K_d about 10^5), independent of the salinity and of the concentration of PCBs in solution (middle bar). Since the major source of PCBs are the rivers Rhine, Meuse and Scheldt, dissolved PCB concentrations (middle bar) are high at low salinities (right bar) and low at high salinities.

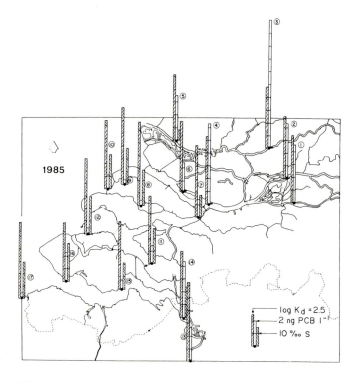

Fig. 5.7. Log K_ds of ΣPCBs (sediment/water; left bars), dissolved ΣPCBs (middle bar) and salinity (right bar) in the Dutch delta region.

The analysis of PCB congeners individually in the dissolved compartment water requires a very sophisticated extraction technique. In open sea, where concentrations are low, at least 50 l water must be sampled with a non-contaminated sampler or pump, and stored in a non-contaminated stainless steel drum. Organochlorines are subsequently pressed into a continuous counter-current extractor according to the technique as used by various scientists (Duursma et al. 1986; 1989; 1991; Duinker et al. 1988).

After specific clean-up, separation and concentration procedures of the extractant (f.e. hexane or pentane), the PCBs are analyzed by gas-chromatography, using capillary columns and an Electron-Capture Detector. PCBs on PM or other solid matter are extracted by Soxleth percolation. For the determination of the K_d it is essential to have both concentrations given in pg or $\mu g/g$ PM and water. The amount of g PM/kg sea water is therefore required.

Since the level of PCBs in PM or surface sediments are much easier to be determined than that in sea water, these are often the only data available. The logical error which is often made that PCBs are therefore mainly attached to PM and sediments (apart from being accumulated in organisms) and that their distribution is primarily guided by transport of PM and sedimentary material. As shown earlier in section 2.3.2.2. (see also Fig. 2.13A and B), this depends on the percentage of PCBs in solution, which is high at low PM concentrations even for high K_ds.

5.2.2 DDT and PCB behaviour in organisms

5.2.2.1 Mussel Watch

5.2.2.1.1 General background

For many decades, mussels have been used as standard reference organisms to determine environmental contamination levels across virtually all marine habitats on the globe. Environmental monitoring is coordinated by the International Mussel Watch programme, which was set up to standardize procedures and intercalibrate results.

The idea behind the programme is that:
- Mussels of various species can be sampled worldwide.
- Mussels, being filter feeders, accumulate metals and organochlorines from water, plankton and particulate matter.
- Accumulation is assumed to give average contamination levels of an environment, in spite of fluctuating concentrations (f.e. due to tides).

- Analysis techniques for mussel tissue are relatively simple and widely known.

However, some potential major problems exist with some of these assumptions. For example, are PCB concentrations, found in mussels, always proportional to those in the water and sediment compartments of the environment?

For the mussel *Mytilus edulis* Duursma et al. (1983) determined that:
- The biological half-time of loss of various organochlorines differ. These are: 6-CB: 7 days; 5-CB: 15 days, Dieldrin: 159 days and alpha and γ-HCH: 127 and 322 days.
- There was a significant correlation between contaminant concentrations (5-CB, 6-CB and ΣPCBs (50% chlorinated)) in mussel lipid and salinity. During subsequent sampling it was found that the salinity was proportional to the PCB concentrations at least in the water compartment (Fig. 5.7), which indicates a proportionality between PCB concentrations in water and lipids of the mussels. This correlation was, however, only valid for data obtained over a period of 2 years for Dieldrin and p,p'DDE, and ½ year for 5-CB, 6-CB, p,p'DDD and ΣPCBs.

Conclusion: The interpretation of Mussel Watch data has to be carried out with great care. Uptake and loss processes must be known which cause the supposed apparent equilibrium between the concentrations in the mussels and those in the water compartment.

5.2.2.1.2 International Mussel Watch Project

The world-wide Mussel Watch investigation is directed by the International Mussel Watch (IMW) Committee and coordinated and administered by the Project Secretariat office at the Coastal Research Center of the Woods Hole Oceanographic Institution. Under the auspices of the International Mussel Watch, the Marine Environment Laboratory, IAEA-MEL, Monaco and the Texas A&M University Geochemical and Environmental Research Group (GERG) are responsible for sample analysis and quality control.

ΣDDT concentrations from 1990-1992 (IMW 1994; Sericano et al. 1995) in mussel samples from South and Central America (Fig. 5.8) range from 10 to 10,000 μg/kg lipid with an average of 839 μg/kg lipid (Fig. 5.9A). This average value, and a K_d (μg/kg lipid : μg/kg water) of approximately 10^6, gives ΣDDT concentrations in water of 0.84 ng ΣDDT/l.

The length of coastline sampled was about 12,000 km, and assuming DDT content is present in a coastal volume of 100 km wide and 100 m depth, a volume of $1.2 \times 10^{13}\,m^3$ is calculated.

Fig. 5.8. Mussel Watch sampling stations for ΣDDT and ΣPCB (IMW 1994).

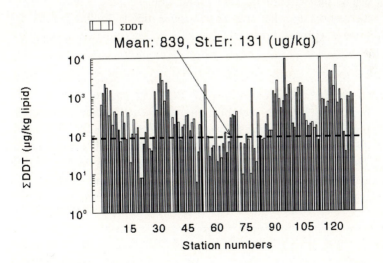

Fig. 5.9A. ΣDDT concentrations in bi-valves of South and Central America.

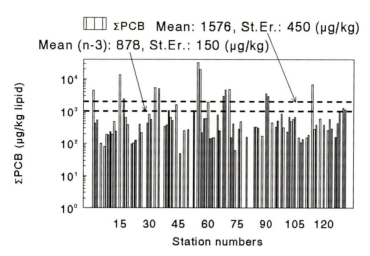

Fig. 5.9B. ΣPCB concentrations in mussels from South and Central America. In calculating the mean (n-3), concentrations above 10,000 μg/kg have been discarded.

The total amount of dissolved ΣDDT present would then be 10.1 ton ΣDDT. A similar value of 10.5 ton is found for ΣPCB (Fig. 5.9B) discarding the three highest ΣPCB concentrations (above 10^4 μg/kg lipid).

This rough calculation shows the utility of the Mussel Watch analysis, even for regions where the amount of ΣDDT and ΣPCB in the environment is negligible with respect to what was previously applied. Similar results are found for Indonesia with respect to DDT contamination (mentioned in section 5.4).

Exercise 5.1. Suppose concentrations of PCBs in mussels, water and (lower) atmosphere are in equilibrium with each other and their concentrations in mussels are on average 300 μg/kg dry weight. The lipid content of the mussels is 3% (/dry weight) and the K_d (lipid/water; kg/kg) is 0.5×10^6. Use as K_d (water-air; m^3/m^3) the value given in Fig. 7.20B (= 122) and calculate the PCB concentration in the lower atmospheric layer (1 atm.). Compare the result with the PCB concentrations in air of Fig. 7.20B.

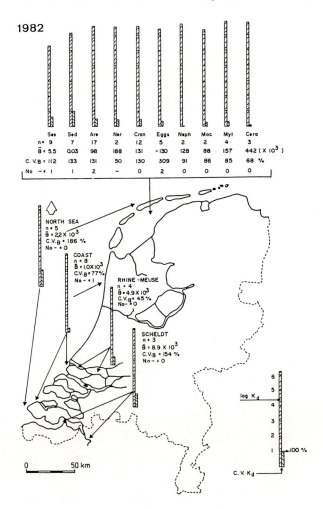

Fig. 5.10. Above: Log K_ds (g/g) of ΣPCBs (lipid/water; left bars) of various organisms, the K_ds calculated from unpublished data of D.Ten Hulscher and P.J.A.Tolsma (cf. Duursma et al. 1986). where: Ses = PM, Sed = sediment, Are = *Arenicola marina*, Ner = *Nereis virens*, Cran = *Crangon crangon*, Eggs = eggs of shrimp, Neph = *Nephtys hombergii*, Mac = *Macoma balthica*, Myt = *Mytilus edulis* and Cer = *Cerastoderma edule*. Below: Log K_ds (g/g) for particulate matter (right bar coefficient variation c.v. of K_d in %).

5.2.2.2 K_ds for organisms

In 1982 a German scientist (Schneider, 1982) posed the hypothesis that organochlorines in marine organisms were in apparent equilibrium with their environmental concentrations in the water compartment, at least when concentrations in organisms were determined on the basis of phospho-free lipids.

This implies that, although there is uptake of organochlorines through food ingestion, the ultimate level of contamination is independent of the food chain. Calculated on lipid basis, the levels of DDT and PCB should be equal for all gill-breathing organisms occurring in the same region.

Fig. 5.10 presents K_d values for ΣPCBs (on lipid basis) of a number of benthic organisms (see figure caption for list of organisms). The log K_ds are equal at the level of 6.5 ± 0.2.

The findings, as given in Fig. 5.12A and B for mussels, confirm this constancy for individual PCB congeners, since the standard error for K_d is very low over a period of 1.5 year. There is similar evidence for some congeners. The level of contamination is given in Fig. 5.11 showing temporal and spacial patchiness.

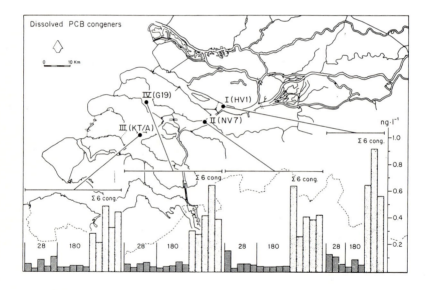

Fig. 5.11. Map of stations (Fig. 5.12A and B) and concentrations of dissolved PCB congeners during five seasons (1985/86) (Duursma et al. 1989).

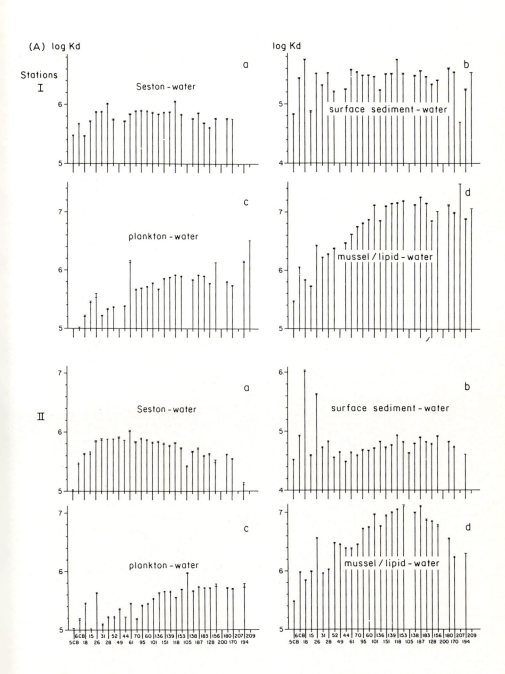

5.12A. Average K_ds, including standard errors, for 5CB, 6CB and PCB congeners of PM (a), surface sediment (b), plankton (c) and mussels (d) at stations I and II (see Fig. 5.11).

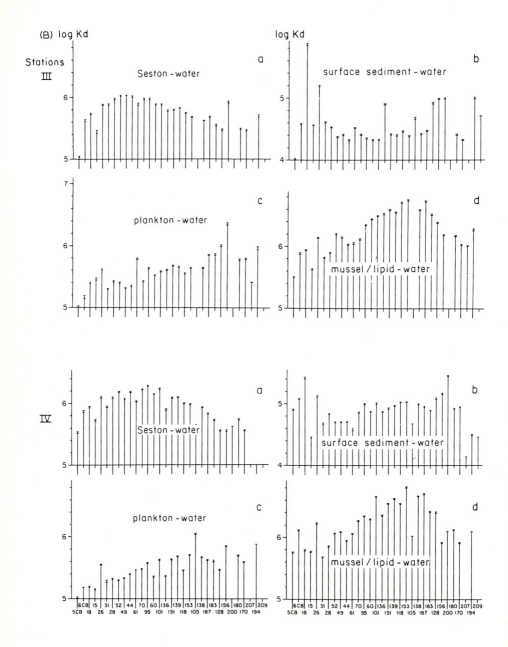

Fig. 5.12B. Ibid as Fig. 5.12.A at stations III and IV of Fig. 5.11.

5.3 The eel case

5.3.1 Introduction

The European eel *Anguilla anguilla* lives most of its life in lakes and streams, but migrates as adult 'silver' eel to the open Atlantic to reproduce. In particular eel living in the Rhine are exposed to contamination by polychlorinated biphenyls (PCBs), which are accumulated in the lipid fraction of these organisms during their stay in contaminated freshwater systems. Research was carried out to answer the question regarding to what extent this contamination continues to be present during their migration in the open sea.

Although the specific partitioning of PCB congeners between water and lipids has not yet been demonstrated for eel, the equal partition principle suggests that loss of PCB congeners should occur when eels migrate from freshwater systems to the sea, where lower dissolved PCB concentrations occur.

It is known, that during migration to their sites of reproduction, eels fast and derive energy from their reserve of lipids and organs (such as their alimentary track). Fasting may have a profound effect on their metabolism and consequently on the PCB congener loss processes from lipid to water compartments. The dominant question is, which? Shifts may also occur in the spectra of specific PCB congeners, indicating specific metabolic activity on the PCB congeners.

5.3.2 A simulation experiment

A simulation experiment was set up in the aquarium of the Centre of Estuarine and Marine Research (CEMO), Yerseke, Netherlands, where a number of female and male migrating adult silver eels, captured in the Rhine river branch the Waal, were kept in running seawater under starvation conditions for 91 days. Eels were periodically sampled for PCB analysis, in order to determine how starvation affects PCB congener spectra and partitioning between the eel lipid and water compartment (Duursma et al. 1991). A group of 21 females and one of 57 males, with a mean length of 60 $\pm$ 10 cm and 35 $\pm$ 10 cm, respectively, were placed in flowing seawater containing low concentrations PCBs. This sea water, provided from reservoirs of an outdoor aquarium was pumped through rectangular aquaria, forcing the eels to swim against the current.

Swimming and the metabolic maintenance of the organism requires energy from the lipid reservoir. It is known that fasting eels, even under conditions of forced (slow) swimming, can last some time without losing significant weight (Boëtius and Boëtius 1967).

Weight loss was estimated by applying the commonly used allometric equation, in combination with an exponential increase with temperature (Fonds et al. 1989):

$$M = e^a \cdot e^{qT} \cdot W^b \qquad (5.8)$$

which relates oxygen consumption M (mg $O_2.d^{-1}$) to body weight W (g) and temperature T (°C), while a, q and b are measured parameters. The parameter q can be converted into a Q_{10} (relative increase in oxygen consumption with a 10° C increase in temperature) with the expression:

$$Q_{10} = e^{10q} \qquad (5.9)$$

Graphical data from Degani et al. (1989) were used to estimate the parameters a (-3.80 ± 0.30), q (0.12 ± 0.012) and b (0.69 ± 0.030) by least squares regression of log M on T and log W (Fig. 5.13).

Table 5.1 shows the calculated weight loss of male and female eels at 5°C. Conversion factors of 13.55 J to 1 mg O_2 and 0.1 mg fresh weight of fish to 1 J were used. For comparison see also Table 5.2.

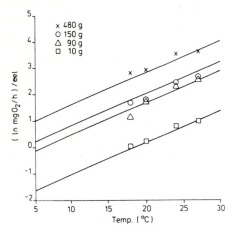

Fig. 5.13. Fit of O_2 consumption of four size classes of eel at different temperatures, based on data of Degani et al. (1989).

Table 5.1. Calculated loss of weight for 91 day fasting eels at low-speed swimming at 5 °C temperature.

	Average Weight (g)	O$_2$ consumption (*) mg O$_2$ d^{-1}	Energy gain J d^{-1}	% weight loss for 91 days (**)
Female	463	64.3 40.2-102.7	871	1.53
Male	83	20.0 12.6-31.5	270	2.56

(*) 95% confidence interval; (**) cumulative %

5.3.3 Discussion

5.3.3.1 Weight, lipid, and water content of eel

Because stored energy reserve was not taken into account for equation (5.8), oxygen consumption was simply related to dry weight. Therefore, this equation is strongly empirical. The loss weight % for 91 days, shown in Table 5.1 support the expectation that little loss of weight will occur in eels held at 5°C and at low swimming activities.

  This is also in agreement with the starvation experiment for male silver eels of Boëtius and Boëtius (1967). Over an experimental starvation period of one year, they measured a loss of initial body weight of 29.2 to 34.7% , depending on the original size of the eel (Table 5.2).

Table 5.2. Loss of weight of fasting male eels over one year at a mean temperature of 13.5 °C (data from Boëtius and Boëtius 1967) compared to a calculated loss of weight on the basis of equation (5.3).

Length (cm) male eels	Weight W (g)	O$_2$ consumed mg O$_2$ d^{-1}	Energy gain J d^{-1}	Boëtius %/yr[a]	% weight loss /yr[b]
33	56.7	43.3	587	34.7	33.2
37	81.2	55.4	750	31.2	30.0
41	105.6	66.2	897	29.2	27.9

[a] determined; [b] calculated

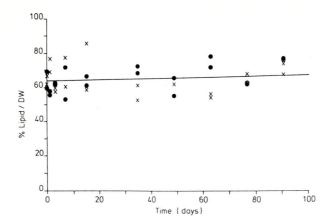

Fig. 5.14. Lipid (fat) in % of dry weight of starving eels. Dots represent males, crosses females. For n = 38 the average lipid content is 64.5 ± 7.7%.

The mean temperature for this period was 13.5 °C. Applying equation (5.9), the theoretical values are calculated and compared to the experimental results. As shown in Table 5.2, the determined loss of body weight was close to what is predicted theoretically.

The observed constant level of lipids per dry body weight (Fig. 5.14) confirms the theoretical low energy utilization as expressed in $J.d^{-1}$ of $\%.d^{-1}$ of body weight (Table 5.1). Calculating this parameter for an average exposure time of 33 days, gives an estimated consumption of 1.0 % of body weight for female eels and of 1.7% for males for the whole experiment of 1-91 days.

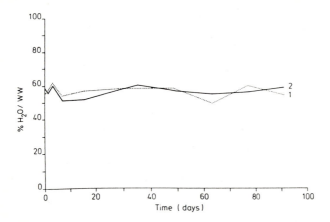

Fig. 5.15. Water content of female (1) and male (2) in % of wet weight of starved eels. The all mean is 56.1 ± 3.2%.

Compensation for loss of lipids by replacement with water is possible in principle, as Boëtius and Boëtius (1967) found for eel with stimulated gonad development. However, the water content, based on wet weight determinations, remains practically constant, except for some 'noise' at time 0 to 3 days, probably due to the osmotic shock by transfer of the eels from freshwater to seawater (Fig. 5.15).

5.3.3.2 Age and condition of organs

Since the determination eel ages on the basis of number of rings in otoliths is still a matter of discussion among fishery biologists, and assuming one ring represents one season to one year, it could be only tentatively concluded that eel ages ranged between 4-12 years for female eel and between 4-14 years for male eel.

The data on the organs, heart, liver and spleen, demonstrated that within the limitations of the methodology, no differences among these organs were observed throughout the experiment. Nor could the sexual maturation of gonads be observed.

5.3.3.3 Partitioning of reference congener 138

The hypothesis for this study was that PCB-congener concentrations in eel as they migrate to the open sea should ultimately adapt to the principle of constant partitioning between dissolved PCBs in the water compartment, and those accumulated in the lipid compartment of aquatic organisms. If the hypothesis is correct, the three-month eel experiment would possibly also indicate a loss of PCB-congeners from eel to water to a degree that the partition coefficients would be similar to those existing earlier at time 0 (Rhine conditions).

The results of this study show that a constant partitioning coefficient Kd (defined as the ratio of PCB in lipid, given in $\mu g.g^{-1}$ and dissolved PCB in $pg.l^{-1}$ $\{x\ 10^3\}$ and expressed in $ml.g^{-1}$) with time is not found for the selected reference congener 138 in eel (Fig. 5.16). There is an expected increase from time $t = 0$ to $t = 1$ day when eel is taken from freshwater to seawater, and a maximum value at time 49 and 63 days, due to the lower concentrations in water and not to higher concentrations in the lipid compartment. From 77 days on, the original Kd values of $t = 1$ to 35 days are being attained again, of the order of magnitude of 1 to 2×10^7 $(ml.g^{-1})$. Over the whole period from 1 to 91 days, there was practically no PCB loss from the eels.

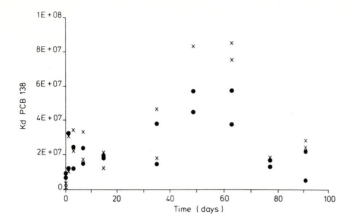

Fig. 5.16. K_ds of the major present PCB congener No. 138 for starving eels. Dots are for females, crosses for males.

The average initial K_d value for PCB-138 of 10^6-10^7 (before starvation; t = 0 days) agrees approximately with the values for K_d found in the same region for the partitioning between water and mussels (*Mytilus edulis*) *in situ* living (Duursma et al. 1989). The 'metabolic' partition was probably effective, becoming low from the moment starvation started.

The hypothesis is that low-metabolic activity of eel might hamper further exchange of PCBs between the sub-compartments lipid and blood. There is no reason to assume that exchange between blood and the surrounding water compartment through the gills changes. Nonetheless, because of low PCB concentrations in blood (which still has to be confirmed), this exchange is also negligible.

5.3.3.4 Behaviour and spectra of other congeners

Irrespective of the partitioning process between water and organism, there is also the relative behaviour of congeners with respect to each other. The abundantly present PCB congener 138 was for the purpose of studying this behaviour chosen as the arbitrary reference congener. As demonstrated by Boon et al. (1989; 1992; 1994), changes with time occur in the spectra of the PCB congeners of aquatic and non-aquatic organisms. These are due to metabolic processes in organisms and are particularly effective for lower-chlorinated and lower numbered congeners.

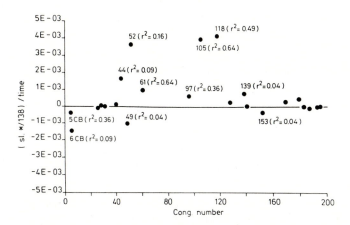

Fig. 5.17. Plot of slopes (sl) of linear regressions of the ratio of [different PCB congeners (*)]/[congener No. 138] with time, for different congener No.

In order to investigate this process for eels exposed to starvation conditions, concentrations of all congeners have been calculated relative to those of congener No. 138. Ratios so obtained were plotted as a function of the time and the linear regression slopes calculated. These slopes are again given in Fig. 5.17 as a function of the congener number, including (in brackets) the correlation coefficient (r^2) for those slopes differing from zero.

The zero line counts for No. 138. Above this there is an increase with time in concentration relative to No. 138; below is a relative decrease. Only a few congeners have significantly increased concentrations relative to that of congener No. 138. These are the numbers 105, 2,3,3',4,4'-pentachlorobiphenyl and 118, 2,3',4,4',5-pentachlorobiphenyl, and to a lesser extent No. 61, 2,3,4,5-tetrachlorobiphenyl.

The increase of congeners 105 and 118 cannot have been caused by simple exchange with the dissolved compartment, since their concentrations in water followed those of congener 138. Hence the relative increases in eel of the congeners 105 and 118 must have another cause, which may be a low-level metabolic one, with stimulated active uptake from water.

The increase of these congeners in eel is in contradiction with the above-mentioned results found by Boon et al. (1992, 1994), who demonstrated metabolic loss of low-numbered congeners. Starvation may have caused different reactions than those observed under normal nourishment conditions.

Both congener 105 and 118 are toxic components, with an *in vivo* ED-50 inhibition of body weight gain of 750 and 1120 μmol/kg, respectively, in immature Wistar rats (Goldstein and Safe 1989). These values are an order of magnitude higher (which means lower toxic effect) than for other PCBs given by these authors. The same is valid for other toxic effects on thymic atrophy and Benzo[alpha]pyrene hydroxylase activities.

5.3.3.5 Exchange and metabolism

The experiment showed that fasting eels do not lose PCBs dissolved in their 'lipid sub-compartment' to sea water. This resembles a fact presented in section 2.3.3.1 for starved winkles (*Littorina littorea* L.), previously contaminated with the radionuclide ^{95m}Tc (Fig. 2.17). The only plausible hypothesis put forth is that feeding is essential for maintaining in aquatic organisms an active exchange process of contaminants between their different sub-compartments, blood, tissue and lipids. Probably the factors k_b and k_l of Fig. 2.16B converge on zero under starvation conditions.

5.4 Pesticide use and contamination with Tambak (Indonesian brackish-water) aquaculture

A few remarks should be devoted to studies on pesticides (contamination and application) in tropical areas. Fish harvest in coastal brackish water ponds of S.E. Asia is several million tons annually, primarily involving *Chanos chanos* (milkfish), *Tilapia mossambica* and *Penaeus spp* (large shrimp).

Brackish-water aquaculture in coastal ponds (e.g. Tambaks in Indonesia), focuses mainly on herbiphorous species like milkfish. Hence competition from other species (fish, mosquito larvae and snails) must be avoided. Pesticides are only applied for this purpose, when they are harmless to the cultured species and man, and retained low-levels at the time of harvest.

Pesticide analysis in remote tropical areas is complicated by a number of problems:
- Heat, resulting in evaporation of chemical extractants like ether.
- Unstable electricity, resulting in breakdown of Gas Chromatography (GC) equipment.
- Sampling and conservation of organisms in remote areas, without deep freezing.
- Financial limitations, resulting in difficulty in maintaining stocks of essential laboratory materials.

These problems require unorthodox often ingenious simple solutions and it is therefore necessary to improvise considerably with certain sampling strategies and extracting techniques. A lower precision than otherwise would be tolerated must be accepted for routine methods. Several environmental questions are so urgent, that an answer with 70% confidence might be completely satisfactory for making policy decisions (f.e. with the application of pesticides to eradicate wildfish from ponds previous to stocking these ponds with fingerlings).

An example of a simplification of sampling and analysis follows: A few grams of material (sediment or fish tissue), sampled or cut with a non-contaminated knife, are placed in the field in a tube of known weight containing 5 ml of hexane. When closed, the sample is automatically preserved and extraction of organochlorines begins. In the laboratory, the tube is weighed again, the tissue crushed with a clean glass rod, and after one day 1 ml of hexane is removed for further analysis, such as clean-up with a few drops of concentrated H_2SO_4, centrifuging and GC analysis of 10 μl of the supernatant hexane. This method should be calibrated on the standard high-level method with Soxhlet extraction of freeze-dried tissue and sophisticated clean-up procedures, well described in the literature.

5.4.1 Environmental contamination in the tropics

5.4.1.1 ΣDDT

In 1975, Indonesia still used a considerable amount of DDT due to the anti malaria program. For their anti-malaria program in Indonesia the production in 1974 was still 24,000 ton/yr. Nevertheless it was surprising to find only low concentrations of DDT and its metabolites in marine species, which was verified for a great number of coastal fish samples obtained from the whole Indonesian archipelago (Duursma 1976). Concentrations in sediments ranged from 0.5-28 ng/g (dry wt.) and for coastal-zone fish species: from 4-500 ng/g (wet wt.). Assuming a dry/wet ration of 10% and 5.2% (standard error 0.26%) lipid/dry wt. (IMW 1994), concentrations for fish species range from 40 - 5000 ng/g dry wt. and 770 - 9.6x10^4 μg/kg lipid, which was equal or higher than those found during the 1990-1992 International Mussel-Watch programme of Central and South America (Fig. 5.9A).

The fact that metabolites of DDT (DDE, DDMU) often dominated DDT in the samples (low DDT/ΣDDT ratio), emphasises the relatively low persistence of DDT in tropical regions. Only in exposed regions did the ratio DDT/ΣDDT approached 0.9.

5.4.1.2 Application of pesticides in aquaculture

Tests have been performed with low persistence, highly toxic organo-phosphorus pesticides, such as Diazinon, in order to determine accumulation and loss rates in various aquatic organisms (Fig. 5.18). A rapid uptake of Diazinon is apparent in Tilapia within the first day, arriving at a K_d (μg/g wet weight : μg/g water) of 5, increasing to 150 at the 5th day. Loss is equally rapid with a biological half-time $t_{1/2}$ between 10 and 36 hours for Tilapia and 7 hours for Milkfish.

During a field test in a pond with the now forbidden Thiodan (Endosulfan) pesticide was applied for the killing of wildfish. Thiodan rapidly accumulated in the liver of several wildfish species during the short period of killing. The same was determined for DDT and its metabolites DDD, DDE and DDMU.

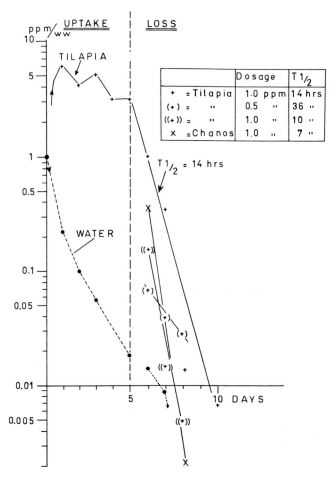

Fig. 5.18. Tests of uptake by and loss of one organophosphorus pesticide, Diazinon for milkfish (*Chanos chanos*) and *Tilapia mosambica*.

Such a rapid uptake of ΣDDT by dying species is difficult to explain. Probably healthy fish keep the DDT, existing in sublethal concentrations at an equilibrium level (as stated earlier above), but when being under stress and dying, excretion is blocked earlier than uptake. This finding remains speculative and requires further detailed studies. However, it indicates that synergistic effects are not simply added together, but often are more subtle and complicated with certain feedbacks.

5.5 Possible DDT and PCB partitioning between air and man

The presence of high levels of organochlorines in terrestrial organisms, such as man, has often surprised investigators. ΣDDT is still today found in low concentrations in the human body (Fig. 5.19). PCBs can also occur in reasonably high concentrations in mother's milk (Jong et al. 1994), where PCBs lack to be present in common food sources of the tested women (Fig. 5.20 and Table 5.3).

Table 5.3. PCB concentration in mother's milk in 1982 (Nieuwenhuize and van Liere, 1982) and 1971-1974 (Anonymous 1980).

No. of mother's milk samples	ΣPCBs μg/g lipid	p,p'DDE μg/g lipid	Age of baby (months)	Vegetarian
185 (1982)	2.2	1.5	0.5	no
186 "	1.0	0.06	0.5	no
236 "	2.4	1.2	3	no
237 "	1.4	0.83	7	yes
263 "	4.4	0.90	5	no
1970-1974	2.7 (average)			
Cow (2 yrs) milk	n.d.	0.03		
Cow (7 yrs) milk	n.d	0.03		
Milk from shop	n.d.	n.d.		
Margarine (n=9)	n.d.	0.005-0.019 0.021-0.053		
Butter (n=2)	n.d.			

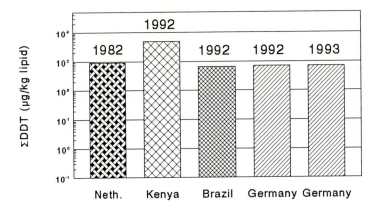

Fig. 5.19. Average ΣDDT concentrations in mother's milk as sampled in the Netherlands, Kenya, Brazil and Germany from 1982-1993, unexposed to recent DDT application.

Investigations of various food, containing lipids, did not reveal any source of PCBs. Also drinking water is usually not contaminated by PCBs. What are then the other uptake pathways, eliminating for now ingestion through food? The most likely remaining source is the atmosphere. In case this would be true, contamination of people should be a wide-spread phenomenon and not only of populations closed to pollution sources.

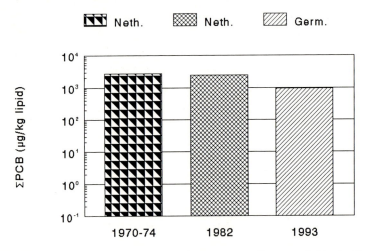

Fig. 5.20. Average ΣPCB concentrations in mother's milk as sampled in the Netherlands and Germany. Anon. (1980), Nieuwenhuize and van Liere (1982), Brunn (1992).

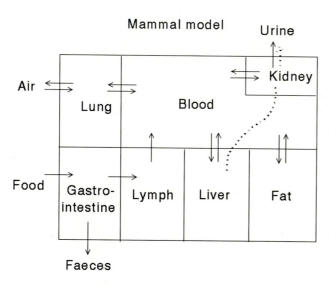

Fig. 5.21. Mammal model based on fish model given in Fig. 2.15.

Converting the model for fish (Fig. 2.15) into a model for mammals (Fig. 5.21), and using the data of the same period for ΣPCBs in rivers (Fig. 5.7) of 5-10 ng/l as a source for atmospheric PCBs, then mammals will inhale these PCBs in such quantities that the concentrations in their blood may reach the same values as observed in the river water, supposing the equal solubility of PCBs in both liquids. Again using earlier Kds for accumulation of PCBs from water to lipids of aquatic organisms of 0.5×10^6, this results in:

$$K_d = 10^6 = \frac{X \ \mu g \ PCBs/g \ lipid}{[5-10] \cdot 0.5 \cdot 10^{-6} \mu g/g \ blood} \qquad (5.10)$$

$$X = [2.5-5] \ \mu g \ PCBs/g \ lipid \ (or \ 2500-5000 \ \mu g/kg) \qquad (5.11)$$

The unknown value X (equation 5.11) is of the same order of magnitude as the level of PCBs in mother milk (Table 5.3). Nevertheless, this agreement is full of potential confounding factors, since there is no definitive proof that an equilibrium exists between PCBs dissolved in natural waters and the lower atmosphere. However, knowing the rapidity of accumulation of certain organic solvents with high partial pressure and volatile toxins like oxine in blood and other organs, it is not inconceivable that PCBs might show a similar partition over much longer time scales.

The same contamination pathway may exist for DDT. As described in section 5.2.2, the DDT concentrations in mussels around South and Central America were 839 μg/kg fat (Fig. 5.8A), which is of the same level of ΣDDT concentrations in mother's milk (Fig. 5.19 and 5.20). This agreement is so striking, that some common cause must be expected, or at least discussed. The hypothesis is that an apparent equilibrium exists between ΣDDT in human lipids, ΣDDT in the atmosphere, ΣDDT in water and thus with ΣDDT in lipids of mussels.

ΣDDT in human lipids was compared with data obtained in Nairobi (Kenya), where DDT concentrations were determined in the human fat, the serum and mother's milk of women giving birth by Caesarean operation (Fig. 5.22).

Concentrations observed in human fat are similar to those in human milk fat, where the partition coefficient K_d between ΣDDT in fat and serum ranges between 477 and 1963 (g/g). This is lower than K_ds between fat and water, probably because of a low fat content of serum.

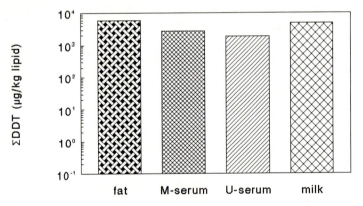

Fig. 5.22. ΣDDT in women (n = 41) as determined in fat, serum and mother's milk in Nairobi, Kenya (Kanja et al. 1992).

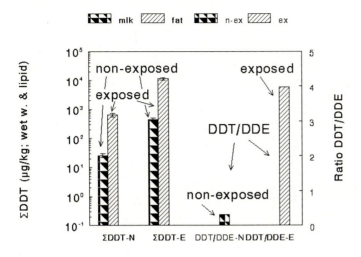

Fig. 5.23. DDTs as determined in mother's milk of two regions in Region Riberao Petro, Brazil, where human populations are exposed to DDT application and in a control region where people are supposed to be non-exposed.

Concerning concentrations of ΣDDT in mother's milk in Brazil, Matuo et al. (1992) made a distinction between women not exposed to DDT (n = 31) and those being occupationally exposed (n = 7). As can be seen from Fig. 5.23, there is a clear difference in ΣDDT concentration in the two graphs. It is also interesting that the DDT/DDE ratio differs similarly for the two groups, indicating that DDT has been converted to its metabolite DDE in the non-exposed women. Their contamination must have been due to long-term exposure and the observed concentration may be in equilibrium with the external ΣDDT concentration in the atmosphere. Whether this fits with a world-wide contamination picture will later be discussed in section 7.3.4.

When demonstrating some kind of global equilibrium between the atmosphere as a source and man, this does not exclude contamination through food. As discussed (see Fig. 5.21), the pathway of intake may occur through food, in spite of the tendency to equilibrate with the environment through breathing. Indeed PCB, p,p'DDT and p,p'DDE concentrations in blood plasma increase in human populations having different fish feeding habits in Sweden (Fig. 5.24). ΣPCBs increased from 970 μg/kg lipid for a non-fish eating group (n = 8) to 1936 μg/kg lipid for a high-fish eating group (n = 11); for information on the different congeners see Asplund (1994) and Asplund et al. (1994).

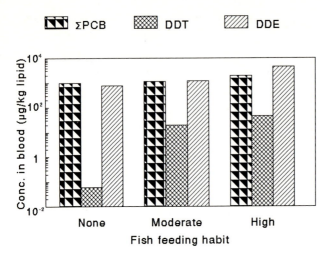

Fig. 5.24. ΣPCB, p,p'DDT and p,p'DDE concentrations in blood serum on groups of men having different fish-feeding habits (Asplund 1994; Asplund et al. 1994).

The same is true for p,p'DDT, increasing in concentration from 0.06 to 45 μg/kg lipid and p,p'DDE, increasing in concentration from 750 to 4,500 μg/kg lipid. It is interesting to note that the ΣDDT (DDT + DDE) concentrations in blood of man in Sweden correspond with those of men determined in Kenya (Fig. 5.22).

Nevertheless both theses of equilibrium-seeking concentrations and accumulation through food may encounter difficulties to explain the differences found by Dewailly et al. (1994) between p,p'DDE concentrations in mother's milk which were statistically different for women given breast-feeding for the first time (328 μg/kg lipid; n=268) and those for a total period of 12 months or more (174 μg/kg; n=43). Perhaps there exist a competition between a slow life-time cumulative uptake of DDE (seeking equilibrium between DDE in the atmosphere) and a more rapid release at the moment of loss of lipids through mothers milk, to be replaced with less contaminated lipids in food. In that case regeneration of lipids in humans, not breast-feeding, must be a slow process.

6 Competitive reactions and effects of conservativity

6.1 Introduction

Conservative mixing of metals in river mouths and estuaries is generally defined as occurring when concentrations of these metals have a linear relationship with salinity (line (b) in Fig. 6.1). When a relationship such as curve (a) exists, one concludes a subsequent addition or production of the dissolved chemical, while in the case of curve (c), a loss by degradation or settling is predicted in the low-salinity region.

Such relationships are, however, more complicated, particularly for estuaries in which sorption and complexing processes occur and where a turbidity maximum is present. Hence, mixing curves of individual metal species must be calculated on the basis of the real behaviour of metals as a function of salinity.

Complexing and sorption - desorption processes are supposed to occur with inorganic and organic complexes and particulate matter respectively. Thus conversions occur from dissolved ionic metal (MET) to organically complexed metal (OMET) on the one hand, and MET and to particulate metal (PMET) on the other hand.

The result may be that for individual metal species (ionic, organo-complexed and particulate), no proportionality exists with salinity, while in fact there is *no* overall net loss or gain of the metal in the estuary. Hence the system as a whole is still 'conservative'. The common definition of conservativity requires therefore insight into the complete system.

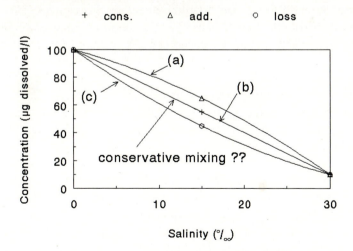

Fig. 6.1. Assumed conservative and non-conservative mixing in estuaries.

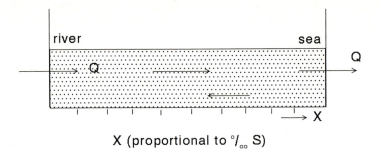

X (proportional to $^o/_{oo}$ S)

Fig. 6.2. Scheme of model estuary. The average salinity at X is proportional to X, whereas width and water depth are constant at any value of X. Q = input equal to output.

6.2 Principles of conservativity

An example is presented (Duursma and Ruardij 1989) where three chemical metal species (ionic, particulate and complexed) are followed through an estuarine mixing zone. In this simplified model, estuarine mixing of freshwater with seawater takes place in a basin of equal depth and width, where the average salinity at distance X is proportional to the length of the estuarine mixing region. The mixing of freshwater and marine particulate matter occurs by inflow of seawater along the bottom from the sea side and freshwater along the surface from the other side, resulting in a turbidity maximum in the brackish zone (Fig. 6.2).

For freshwater, brackish water and seawater the stability factor (K_{st}) for complexation and distribution coefficient (K_d) for sorption may both be constant, at least as far as this model is concerned.

6.3 Model study of competitive processes in an estuary containing a turbidity maximum

6.3.1 Boundary conditions

A model estuary is considered, in which river water mixes with seawater (including a salt wedge), and freshwater particulate matter mixes with marine particulate matter (including a turbidity maximum). The conditions (within tidal variations) are considered to be steady state, while no residual sedimentation or resuspension occurs. The only metal sources are river and sea. Table 6.1 presents the chosen data for concentrations of total dissolved metal (TDMET) and dissolved organic matter (DOM) in river and seawater, suspended particulate matter (PM) and the equilibria constants K_{st} (stability constant) and K_d (distribution coefficient).

6.3.2 Reactions

The reactions of metal with different phases (or compartments) are defined by:

$$
\begin{array}{c}
MET + DOM \underset{}{\overset{K_{st}}{\rightleftharpoons}} OMET \\
\Updownarrow \\
MET + PM \underset{K_d}{\rightleftharpoons} PMET
\end{array}
\qquad (6.1)
$$

where:

$$
K_{st} = \frac{[OMET]}{[MET][DOM]} \qquad (6.2)
$$

having the dimension of [l.meq^{-1}], where (l = litre and meq is milli-equivalent), and

$$
K_d = \frac{[PMET]}{[MET]} \qquad (6.3)
$$

having the dimension [ml.g^{-1}], where ml = millilitre and g = gram. These factors are both assumed to be constant throughout the estuary.

There is supposed to be no DOM production or decomposition, no precipitation or dissolution of DOM and also no residual sedimentation or resuspension of PM. The reactions (6.1) are considered to be fast with respect to the freshwater - seawater mixing process and completely reversible.

6.3.3 Simulation model

The simulation model is based on a one-dimensional mixing and flushing model (Zimmerman, 1976). The model estuary is divided into 15 compartments of equal length and 2 ‰ salinity intervals, through which a transport of metal, organic matter and PM occurs. The steady state distribution of PM concentrations is given in Table 6.1. Only advective transport is assumed to transport particulate matter because diffusive processes are likely to be negligible on the scales concerned. Advective transport is equal to the product of the river discharge and the concentration of PM at 0 salinity. In this way we were able to transport silt (PM) without disturbing the given PM distribution.

A simulation model was used with initial concentrations of the variables MET, OMET and PMET as inputs for the calculation of each compartment. The metal concentrations are initially determined for a hypothetical scenario in the estuary in which no transport of PM occurs. In this situation, the PMET distribution is in equilibrium with the MET distribution, and there is no net exchange between MET and PMET.

Consequently MET + OMET = TDMET (total dissolved metal) should behave conservatively with ‰ S. Recall that there is input to the estuary with subsequent mixing and discharge to the sea, but no loss or gain to the particulate phase in the estuary. This formula follows from these assumptions:

$$TDMET_S = \frac{TDMET_0 \cdot (S_{sea} - S) + TDMET_{sea} \cdot S}{D_{sea}} \tag{6.4}$$

in which S_{sea} is the salinity at sea, $TDMET_0$ the TDMET concentration of the river at 0 salinity and $TDMET_s$ the concentration of TDMET at S salinity.

From (6.2) the initial values for MET and OMET can be calculated:

$$TDMET = OMET + MET = MET \cdot DOM \cdot K_{st} + MET \tag{6.5}$$

or:

$$MET = \frac{TDMET}{1 + DOM \cdot K_{st}} \tag{6.6}$$

and:

$$OMET = \frac{TDMET \cdot DOM \cdot K_{st}}{1 + DOM \cdot K_{st}} \tag{6.7}$$

PMET can now be derived from (6.3) and (6.6):

$$PMET = \frac{TDMET \cdot K_d}{1 + DOM \cdot K_{st}} \tag{6.8}$$

The border conditions for MET, OMET and PMET at 0 salinity are calculated in the same way from $TDMET_o$.

The magnitude of the river discharge (advective transport) results in a residence time of 30 days. The exchange coefficients (EX), which describe the diffusive transport between the compartments of 2 ‰ S, are derived from the salt distribution (Zimmerman 1976):

$$EX(I,I+1) = \frac{Q \cdot (SALT(I) + SALT(I+1)) \cdot 0.5}{SALT(I) - SALT(I+1)} \qquad (6.9)$$

in which Q is the river discharge $(m^3.s^{-1})$ and SALT(I) is the salinity in compartment I.

The model simulation starts with the equilibrium situation with no PM transport and continues until equilibrium with PM transport is reached. This situation was reached after 360 days. On this basis it was decided to apply the model as given.

Table 6.1. Data of parameters and constants.

Parameters	River	Sea		Run			
				1	2	3	4
DOM mg/l	2.0	0.0	$logK_{st}$	2	4	2	4
S ‰	0.0	30.0	$logK_d$	3	3	5	5
TDMET=MET+OMET mg/l	0.004	0.005					

S	0	2	4	6	8	10	12	14	16	18	20	22	24	26	28	30
PM	10	10	14	20	60	50	30	20	15	10	6	6	5	4	4	4

DOM = dissolved organic matter, able to complex Me (mgl^{-1})
Me = metal with MET = metal-ion, OMET = metal-organic, PMET = part. metal
Equiv. weight Me = 30, Equiv. weight DOM = 60
PM = particulate matter (e.g. suspended sediment) (mgl^{-1})
K_{st} = metal-organic stability constant (dimensions $l.meq^{-1}$)
K_d = distribution coefficient (dimensions $ml.g^{-1}$)
S = salinity in ‰

6.3.4 Results and discussion

Four different cases have been calculated with K_{st} and K_d having different values (Table 6.1). The results are given in Fig. 6.3, where the concentrations of MET, OMET and PMET are plotted against the ‰ salinity.

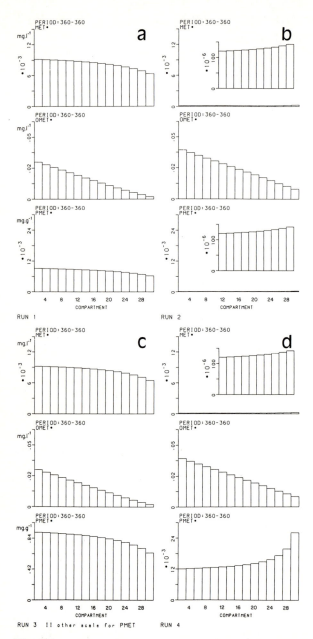

Fig. 6.3. Four runs (Table 6.1) of three metal species: MET, OMET, PMET against salinity, for different K_{st} and K_d values. MET and OMET in mg dissolved Me l^{-1}, PMET in mg Me g^{-1} particulate matter. The data are the results from a 1360-day run of the model. The notations **a**, **b**, **c** and **d** correspond to runs 1, 2, 3 and 4, respectively.

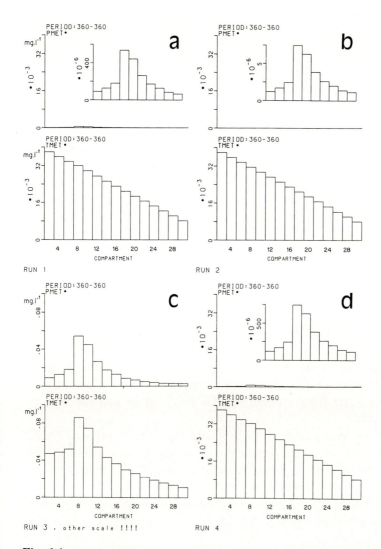

Fig. 6.4. Plot of particulate metal (PMET) and total metal (TMET), all expressed in mg Me l^{-1}. TMET = MET+OMET+PMET. The notations **a**, **b**, **c** and **d** correspond to runs 1, 2, 3 and 4, respectively. For (a), (b) and (d) PMET, see also insets.

At the imposed boundary conditions, the distribution of metal differs in each case over the three phases (compartments) dissolved ionic, dissolved organic complex and particulate. Since DOM is high in freshwater and zero in seawater, there is an apparent proportionality of OMET with ‰ S, while MET and PMET can either decrease or increase with ‰ S, having non-linear correlations. The amount of metal in particulate matter (PMET, in mg.l^{-1}) has a distinct maximum at the turbidity maximum (Fig. 6.4). The distribution of total metal (TMET) depends on the factors K_{st} and K_d.

What can be concluded from this tentative exercise is that apparent linear mixing plots of metal species concentrations with salinity do not imply that metal species will necessarily behave conservatively, but that exchange processes of complexing and sorption must be taken into account.

This conclusion is fundamental, in spite of the fact that the simulation model needs extension with a more realistic transport model for PM and application to an existing estuary.

Irrelevant to this model, but relevant to field conditions is the point whether the factors K_{st} and K_d can be considered to be constant over a complete range of salinity. Significance analyses should reveal to what extent such ranges, at high levels of K_d, are acceptable for simulation modelling of conservative mixing processes.

6.4 Residence times of metals in an estuary

6.4.1 Theory

In many estuaries of the world, bottom sediments are contaminated with heavy metals. Dissolved metals are sorbed to suspended PM, settled to the bottom at the sediment-water interface and migrate by diffusion into the bottom (Meent et al. 1991).

A case study was carried out in the Hollands-Diep - Haringvliet, a Dutch estuary, where the dissolved and sediment-attached metals were analyzed over a number of years. The multiple-box model approach of DiToro and Paquin (1984) was used, which is given for one box in Fig. 6.5.

The mass-balance equation for the total chemical concentration in the water column C_{T1} and sediment C_{T2} is shown by:

$$\frac{dC_{T1}}{dt} = \left(\frac{dC_{T1}}{dt}\right)_{outflow} + \left(\frac{dC_{T1}}{dt}\right)_{diffusion} + \left(\frac{dC_{T1}}{dt}\right)_{particle} + \left(\frac{dC_{T1}}{dt}\right)_{discharge} \quad (6.10)$$

and for the sediment:

$$\frac{dC_{T2}}{dt} = \left(\frac{dC_{T2}}{dt}\right)_{diffusion} + \left(\frac{dC_{T2}}{dt}\right)_{particle} \quad (6.11)$$

Using these definitions the mass balance equations for the water and sediment columns are give by:

$$\frac{dC_{T1}}{dt} = -\frac{Q}{V}C_{T1} - \frac{K_L}{H_1}\left(f_{d1}C_{T1} - f_{d2}C_{T2}\right) - \frac{W_a}{H_1}f_{p1}C_{T1} + \frac{W_T}{V} \quad (6.12)$$

and:

$$\frac{dC_{T2}}{dt} = \frac{K_L}{H_2}\left(f_{d1}C_{T1} - f_{d2}C_{T2}\right) + \frac{W_a}{H_2}f_{p1}C_{T1} - \frac{W_s}{H_2}f_{p2}C_{T2} \quad (6.13)$$

respectively.

The turn-over or residence time of chemicals in the water column were defined by Vrie and Duursma (1986) as:

$$\tau_0 = \frac{M_0}{F_0} \quad (6.14)$$

where M_0 is the amount of material present in the reservoir and F_0 the material passing through the reservoir per unit time. Residence times give an indication of the rate at which material passes through a system in steady state conditions. They do not, however, give reliable indication of the speed a system adjusts itself from one steady-state situation to another when a change in input (for example less contamination) occurs.

It is just this adaption time that is of interest to predicting future behaviour of systems after a change in the rate or magnitude of contaminant inputs. A 50% adaption time ($t_{50\%}$) can be calculated from:

$$C(t_{50\%}) = \frac{C(o) + \lim_{t \to \infty} C(t)}{2} \quad (6.15)$$

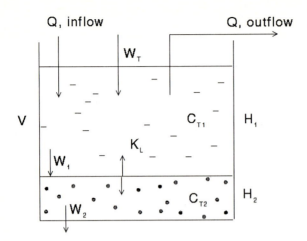

Fig. 6.5. One box model for the determination of residence times of chemicals in the water and sediment column, where: W_T = loading, C_{T1} = total concentration in the water column, C_{T2} = total concentration in the sediment, W_1 = settling, V is volume of water, W_2 = sedimentation, K_L = diffusion-exchange, H_1 = height water column, H_2 = height sediment layer.

6.4.2 Example of river branch

A 3-box model is devised to illustrate the concept of adaption time. Parameters in the model are inflow and outflow of water, bidirectional exchange with a 10 cm bottom layer through settling and diffusion (in and out). The concentrations of the metal Pb can be calculated for the dissolved and sediment compartments for the river branch Hollands Diep-Haringvliet. As the Fig. 6.6 shows, the calculated concentrations matched fairly well with the determined values.

In another example, the turn-over time (residence time) and 50% adaption time of three metals Zn, Cd and Pb were determined in an adjacent river branch (Lake Zoom), using a two box model. As shown in Table 6.2 there is a particularly large difference between the turn-over (residence) time and $t_{50\%}$ for the sediment layer.

The data suggest that it will take 1230 yr either to contaminate 10 cm of the bottom or to have it again completely regenerated (100%), supposing the water over the estuary becomes uncontaminated at t = 0. For 50% of the contamination level it will last only 45 yr for Zn and 69 and 70 yr for Cd and Pb, respectively.

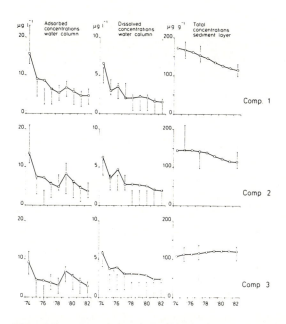

Fig. 6.6. Calculated (o-o) and measured (|) concentrations of Pb in the Hollands Diep-Haringvliet (Netherlands), divided into three compartments (Comp. 1, 2 and 3). The flow direction is towards compartment 3.

Table 6.2. Turn-over and 50% adaption times in Lake Zoom using a two box model.

| | Turn-over times (yrs) | | | | $t_{50\%}$ (yrs) | | | |
| | Water | | Sediment | | Water | | Sediment | |
	Box1	Box2	Box1	Box2	Box1	Box2	Box1	Box2
Zn	42	86	975	1230	18	37	24	45
Cd	42	86	975	1230	34	68	35	69
Pb	39	86	975	1230	128	59	36	70

6.5 Sources and sinks of chemical constituents in estuaries

6.5.1 Introduction

Competitive reactions between sediments and seawater are often responsible for the losses and additions of chemical constituents observed in mixing diagrams (Fig. 6.1). When interpreting mixing diagrams, another aspect which must be considered is whether additional sources or sinks are responsible for the observed increases or decreases in chemical concentration with salinity. The notion of a single channel with inflow at one end and outflow at the opposite end, while useful for instructional purposes is obviously an oversimplification of most natural systems. A more realistic scenario would include multiple source regions supplying dissolved and particle-bound chemical constituents to an estuary. Inflow of varying amounts of freshwater, sediment and chemical constituents is added to estuaries via i) rivers draining geologically distinct source regions, ii) drainage channels located along the margins of estuaries, and iii) overland sheet flow for low-lying areas. While drainage and overland run-off fresh or brackish water supplies are often small, constituent concentrations can be considerably higher than in the estuary itself. Salt marshes and mangrove systems located adjacent to estuaries are particularly effective suppliers of materials to adjacent waters (Carroll et al. 1993; Bollinger and Moore 1993).

This concept is most readily observed in the world's largest estuarine systems where large spatial and temporal scales magnify the influence of exchange processes at the river-ocean interface. An example is the estuarine mixing zone of one of the world's largest river systems, the Ganges-Brahmaputra (G-B). The G-B system is the fourth largest river system in the world in terms of freshwater discharge and first in terms of sediment discharge supplying 1.5 x 10^{12} kg of sediment annually (Milliman and Meade 1983). The distribution and behaviour of the naturally occurring radionuclide ^{226}Ra and its geochemical analogue Ba were investigated to determine sources and sinks for these constituents and to quantify their fluxes to the Bay of Bengal (Fig. 6.7). The G-B rivers are a major source of chemical constituents to the oceans.

6.5.2 The Ganges-Brahmaputra river-estuarine zone

Several large riverine channels supply freshwater to the G-B estuarine zone (Fig. 6.8). The major sediment and freshwater outflow occurs through the Shabazpur channel. Freshwater overflow from the Shabazpur channel is discharged through adjacent channels, i.e. the Hatia, Sandwip and Tetulia channels. The Haringata is a distributary channel which branches from the main stem of the G-B river channel.

The turbidity maximum zone for this estuary extends as far seaward as the 10-m depth contour during low discharge (Barua 1990). Within this zone, suspended sediment concentrations in seawater range from tens to thousands of mg/L, mainly as a result of the resuspension of sediments by tidal currents.

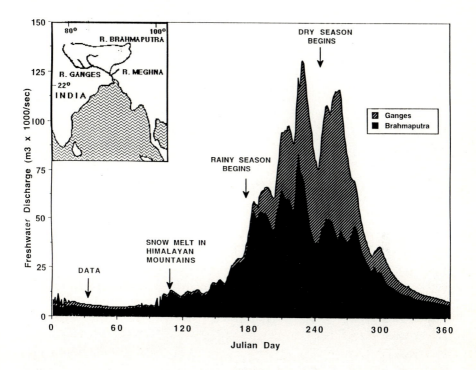

Fig. 6.7. Daily freshwater discharges (in $10^3 m^3/s$) for the Ganges and Brahmaputra Rivers during 1987.

A dominant feature of this estuary is the Sunderband Mangrove Forest located to the west of the major river channel. This region encompasses an area of 2.8×10^4 km^2 and consists of thick deposits of unconsolidated sediments. Eight major tidal channels exchange material and water between the mangrove forest and the coastal ocean. Thus the entire coastal zone interface between freshwater and seawater exhibits a variety of regions whereby the processes of particulate and dissolved matter exchange from land to sea are accomplished.

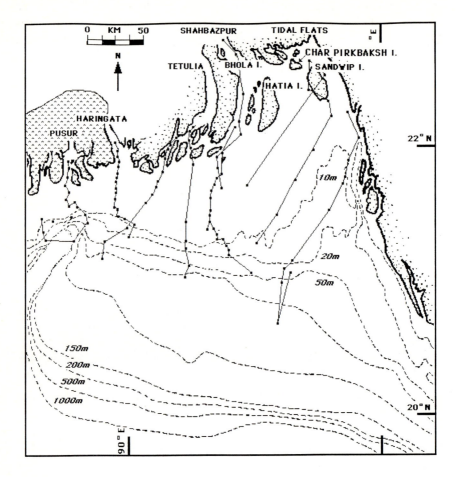

Fig. 6.8. Station locations in the Ganges-Brahmaputra mixing zone during February 1987.

6.5.3 Ba and ^{226}Ra Distributions

In February 1987, ^{226}Ra and Ba samples were collected along transects extending across the salinity gradient from land to sea for the main river outflow channels and a few secondary channels of the G-B system during minimum discharge. Using ^{226}Ra and Ba together, the dominant sediment sources responsible for supplying these constituents to the mixing zone could be identified.

Riverine fluxes of dissolved ^{226}Ra and Ba are generally enhanced in estuarine mixing zones where suspended fluvial sediments desorb radium and barium in exchange for the more abundant cations in seawater (Key et al. 1985; Elsinger and Moore 1980; Li and Chan 1979). [226]Ra is also supplied by sediments deposited in estuaries where [226]Ra produced from the decay of [230]Th escapes by porewater diffusion and sediment mixing processes. However, this process is relatively minor in comparison to desorption of [226]Ra from sediments entering the mixing zone for the first time (Key et al. 1985).

Barium profiles in the river-dominated channels of the mixing zone (Shahbazpur and Haringata) are similar to barium profiles reported previously in estuaries (Li and Chan 1979; Boyle 1976; Edmond et al. 1978). Barium concentrations increase within low-salinity waters (< 6 ‰) and then increase linearly with salinity (Fig. 6.9). These results suggest that the equilibrium distribution of solid and dissolved phases of barium is established shortly after the solids enter the freshwater/seawater transition zone.

In the tidally-dominated channels (Pusur and Sandwip) desorption from freshly supplied river sediments does not explain the relatively high barium concentrations ($\tilde{} $ 450 nmol/L).

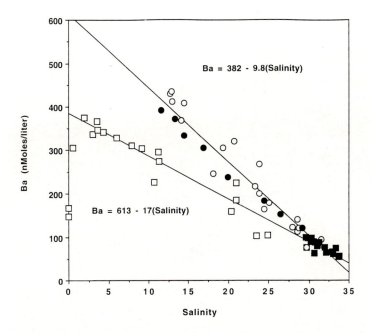

Fig 6.9. Dissolved barium profiles in the Ganges-Brahmaputra mixing zone and the Bay of Bengal (■) during February 1987. The lower linear trend is defined by the river-dominated channels: Shabazpur+Haringata (□). The upper linear trend is defined by the tidal-dominated channels: Sandwip (○) and Pusur (●).

These channels receive little freshwater input during low discharge. The question arises, is there an alternative sedimentary source present in these regions?

The broad curvature of the radium vs. salinity profile for the Shahbazpur and Haringata channels is similar to profiles reported previously for other river mixing zones and suggests that the majority of the observed radium is supplied from suspended sediments entering the mixing zone from the rivers (Fig. 6.10). Like barium, radium activities in the tidally-dominated channels are higher than activities in the Shahbazpur and Haringata channels.

While production and release from mixing zone sediments may contribute to higher ^{226}Ra activities, it does not explain the corresponding increases in barium concentrations. Only desorption can maintain both high concentrations of barium and high activities of radium in the mixing zone (Fig. 6.11). Therefore, barium and radium must both be desorbing from sediments not previously exposed to seawater. The source of these sediments and the corresponding increases in dissolved barium and radium are now described.

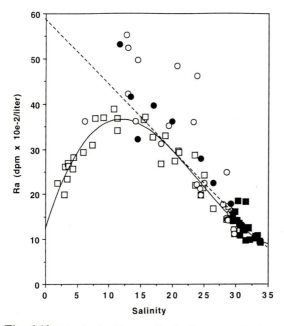

Fig. 6.10. Dissolved radium profiles in the Ganges-Brahmaputra mixing zone during February 1987. Data from the river-dominated channels define the polynomial. Data from the tidal dominated channels lie above the polynomial. The dashed line represents the tangent to the polynomial used to define the apparent endmember concentration for the river-dominated channels. Symbol legend is given in Fig. 6.9.

6.5.4 Sources of Ba and ^{226}Ra

Conditions in the G-B mixing zone are controlled by the pronounced seasonal cycle of freshwater and sediment discharge (Fig. 6.11). As a result of the switch between the southwestern and northeastern monsoons, freshwater and sediment discharges both vary by a factor of approximately 10. In February, freshwater and sediment discharges are generally at their minimum, 7500 m^3/s and 1400 kg/s, respectively in 1987. The flow of both rivers crest during July-August.

During the peak in freshwater discharge, 90% of the annual sediment supply is transported to the mixing zone, while the maximum sediment discharge occurs during August-September (approximately 12 x 10^{14} kg/s; Coleman, 1969). Some of the sediment is trapped in shallow nearshore areas of the coastal zone (Eysink 1983).

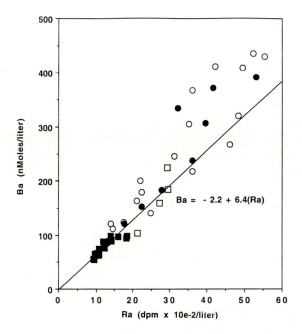

Fig. 6.11. Radium vs. barium in the Ganges-Brahmaputra mixing zone during February 1987. Most of the data from the tidal-dominated channels fall above the linear trend defined by the data from the river-dominated channels. Symbol legend is given in Fig. 6.9.

Inland seawater intrusion is an annual cycle which responds to the seasonal cycle of freshwater discharge. During high discharge when the majority of river sediment is supplied to the mixing zone, increased freshwater input from the rivers shifts the high ionic strength seawater well offshore, away from nearshore sedimentary environments (Eysink 1983).

Because minimal freshwater discharge and maximal seawater intrusion occur each year during February-March, the barium and radium distributions represent the time of near-maximal inland seawater intrusion and near-maximal sediment exposure to seawater.

Based on salinity samples collected throughout the mangrove forest region, seawater (1-2 ‰) extends as far inland as 100 km during minimum discharge (Eysink 1983). In February 1987, salinities at all stations of the tidally-dominated channels were above 10 ‰.

Knowledge of the conditions in the estuary combined with observations on the behaviour of Ba and Ra suggested that in addition to suspended sediments supplied directly from rivers, river sediments deposited during high discharge in mangroves and islands are desorbing barium and radium to seawater during low discharge. The barium and radium from these sediments are likely released in response to the large seasonal seawater intrusion associated with the monsoonal climate and the low-relief of the land at the margin of the Bay of Bengal. This release of radium and barium from deposited sediments in the G-B mixing zone is out-of-phase with the discharge of sediment from the rivers. Seasonal flux estimates in the G-B mixing zone should consider both the direct supply/removal of constituents from river-suspended sediments and the indirect supply/removal from sediment deposits.

6.5.5 Fluxes of Ba and ^{226}Ra

If most of the sediment in nearshore environments is exposed to seawater during the annual seawater intrusion cycle, then knowing the sources and concentrations of barium and radium, annual fluxes of these constituents can be estimated. The total flux of a constituent is equal to the river-suspended sediment flux (F_{SED}) multiplied by the concentration of the constituent desorbed per gram of sediment (C_{DB}), plus the flux of river freshwater (F_{FW}) multiplied by an average concentration of the constituent in river water (Li and Chan 1979) as follows:

$$Flux = \left(D_{(DB)} \cdot F_{(SED)}\right) + \left(C_{(DS)} \cdot F_{(FW)}\right) \qquad (6.16)$$

The resultant flux of radium is 95×10^{13} Bq/yr, and Ba is 53×10^{7} mol/yr (Table 6.3). About 75% of the flux of barium and 95% of the flux of radium from the G-B mixing zone to the Bay of Bengal is supplied by desorption from sediments. Clearly, supplies of both radium and barium from the G-B mixing zone are regulated by the supply of sediments these rivers discharge to the Bay of Bengal.

Table 6.3. Concentrations and fluxes of barium and ^{226}Ra for the G-B river system and mixing zone.

	C_{DB}	F_{SED} $\times 10^{14}$ g/yr	Flux Desorb	C_A	C_{DS}	F_{FW} $\times 10^{14}$ L	Flux Diss.	Flux Total
Ba	255 nmol/g	15	3.8×10^8 mol/l	382 nmol/l	157 nmol/l	10	1.6×10^8 mol/yr	5.4×10^8 mol/yr
^{226}Ra	0.6 Bq/g	15	9×10^{14} Bq/yr	0.58 Bq/l	0.05 Bq/l	10	0.5×10^{14} Bq/yr	9.5×10^{14} Bq/yr

7 Examples of distribution patterns in estuaries and seas

7.1 Budget of plutonium and some other nuclides in an estuary

The distribution of artificially produced radionuclides in suspended matter and sediments was investigated in the Dutch Delta of the main European rivers Rhine, Meuse and Scheldt during 1979-1984 and again in 1986-1991, after the Chernobyl accident (Duursma et al. 1985 and Martin et al. 1994). Potential sources of radionuclides are, besides Chernobyl, fallout nuclides from earlier atmospheric bomb-tests (Fig. 7.1), re-entry and burn-up in the atmosphere of a satellite with a radionuclide energy source in 1964 (RIME 1971), and effluents from either the reprocessing plants of la Hague (France), Sellafield (U.K.) or Mol (Belgium) or nuclear power plants along the fore-mentioned rivers (Table 7.1).

Table 7.1. Nuclear power plants discharging liquid effluents into the rivers of the Dutch Delta (After Luykx and Fraser 1980; 1983).

River	(Country)	Location	Max output capacity (74-'80) in MegaWatt
W. Scheldt	(Nl)	Borssele	450
	(B)	Doel Σ1,2,3,4	2690
Rhine	(Nl)	Dodewaard	52
	(G)	Karlsruhe	52
	"	Obrigheim	328
	"	Biblis ΣA,B	2386
	"	Philippsburg	864
	(F)	Fessenheim Σ1,2	1780
Meuse	(B)	Tihange Σ1,2,3	2770
	(F)	Chooz	305

Fission isotopes were initially supplied to the Delta area from the major atmospheric nuclear weapon testing period, encompassing 1954-1963 (Fig. 7.2). The consequent concentrations of ^{90}Sr (and ^{137}Cs) in the world oceans peaked during the period 1957 until 1967 (Fig. 7.3). Fallout to oceans and continents included the long-lived ^{239}Pu and ^{240}Pu isotopes, which amounted to 400 kCi (1.48×10^{16} Bq) in total.

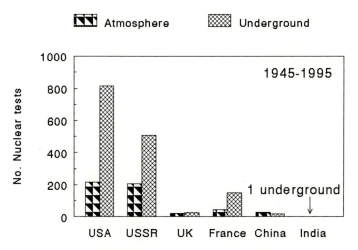

Fig. 7.1. Total known atmospheric and underground nuclear tests between 1945 and 1995 (including 1 Chinese and 2 French 1995 tests) (NRDC 1995).

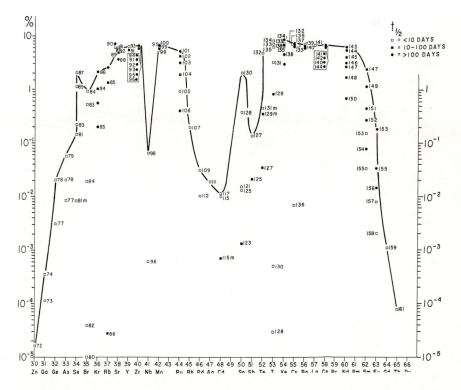

Fig. 7.2. Fission isotope spectrum in % of yield from 100 atoms of ^{235}U, classified in three groups of increasing half-life (Duursma 1972).

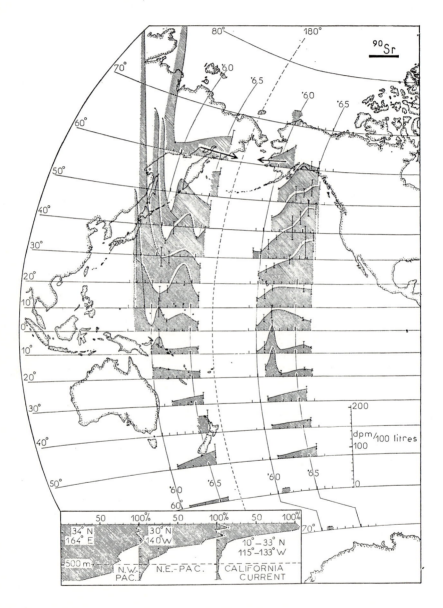

Fig. 7.3. (Left and right) Mean annual ^{90}Sr, or ^{90}Sr calculated from ^{137}Cs, concentrations of surface waters, plotted for 10° bands of latitude. Inserts: some vertical distributions. Data compiled by Volchock et al. (1971) and plotted by Duursma (1972).

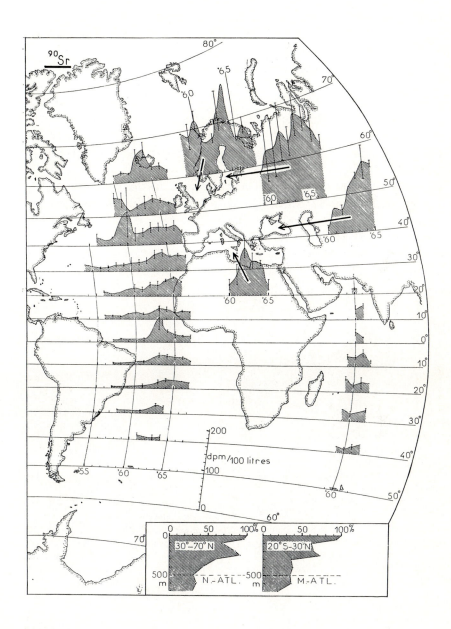

As a consequence of the launching failure of SNAP-9A satellite in 1964, 17 kCi of ^{238}Pu was added to the atmosphere (RIME 1971). This resulted in an activity ratio of ^{238}Pu/$^{239+240}$Pu in fallout averaging 0.04 in the Northern Hemisphere in 1970-1971 (Hardy et al. 1973), and remaining fairly constant afterwards (Thein et al. 1980).

Only nuclear tests, carried out in the atmosphere until 1963, contributed to the world-wide distribution of a great number of fission radionuclides, the spectrum of which being given in Fig. 7.3. Hence, it is possible to calculate from a limited number of determined fallout nuclides in the oceans how much of other fission nuclides should have entered, taking the percentage of each and half-life into account (Duursma 1972). For later fallout data see IAEA (1986a).

It is claimed that little or no nuclides escape to the environment from underground nuclear tests, although leakage of the noble gas nuclides from either Argon, Krypton and Xenon remains possible.

Since the activity ratio of ^{238}Pu/$^{239+240}$Pu in effluents of nuclear fuel reprocessing plants differ from that of atmospheric fallout, this accurately measured ratio became a means to trace sources of contamination. This method has been applied for the Dutch Delta region (Fig. 7.4.).

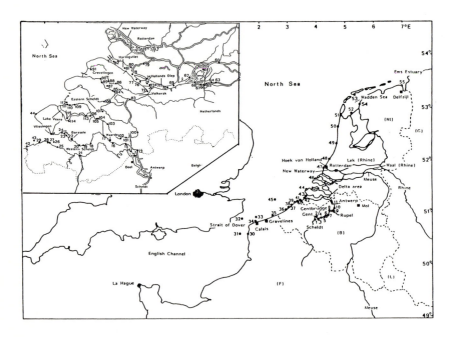

Fig. 7.4. Map of southern North Sea and Dutch Delta.

The $^{238}Pu/^{239+240}Pu$ ratio of various substrates in the Dutch Delta region ranged from the fallout ratio of 0.04 to about 0.4 (Fig. 7.5.). The lower values came from older material in cores and salt marshes, while higher values were evident primarily in recently contaminated material.

For the Scheldt river and Western Scheldt estuary, $^{238}Pu/^{239+240}Pu$ ratio ranged from 0.04 to 0.6 (Fig. 7.6), with maxima values close to the river branch Rupel connecting the Scheldt with the Euro Centre of Mol.

Similar deviation in the $^{238}Pu/^{239+240}Pu$ ratios were found in coastal sediments and mussel flesh of the coast of North-West Europe from the Bay of Biscay until the Elbe mouth in the North Sea (Fig. 7.7). The effluents of the nuclear reprocessing plant of La Hague (France) are clearly sources of ^{238}Pu, while the Scheldt estuary also supplies a contribution.

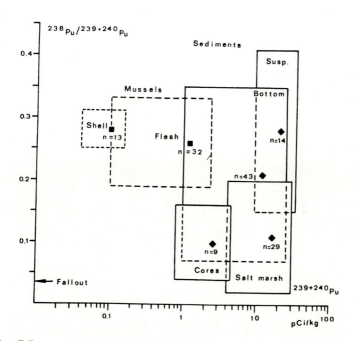

Fig. 7.5. Average data of plutonium isotopic ratios in sediments (until 20 cm depth) and mussels (*Mytilus edulis*) in the Dutch Delta (1979-1984). Dimensions of each box is one standard deviation, n = number of samples analyzed.

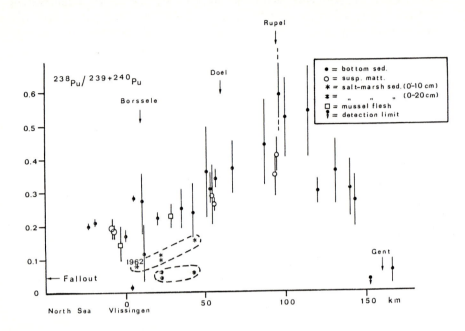

Fig. 7.6. Isotopic ^{238}Pu/$^{239+240}$Pu concentration ratios (by activity) in sediments and mussels of the Western Scheldt Estuary and Scheldt river.

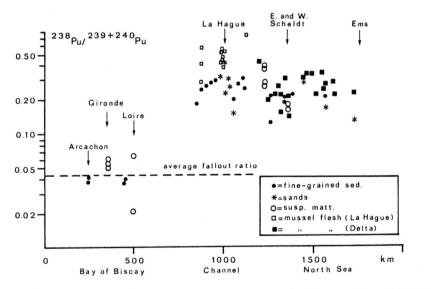

Fig. 7.7. Isotopic ^{238}Pu/$^{239+240}$Pu concentration ratios in coastal sediments and mussel flesh of North-West Europe (1978-1984).

Taking the Western Scheldt as a box model (Fig. 7.8), a rough budget was prepared for both ^{238}Pu and $^{239+240}$Pu for the area between km 50 and km 100. Assumptions were that:

(i) concentrations measured in the surface layer of the bottom deposits and in the suspended matter are representative of the sediments entrapped in this major turbidity maximum zone,

(ii) concentrations are not affected by reworking with older sediments,

(iii) plutonium is fixed rapidly by sediments and

(iv) post-depositional migration in the estuary is negligible.

The conclusion that industry, such as power plants and the centre in Mol, contributed to ^{238}Pu concentrations during the period of investigation, could not be confirmed from their official release figures.

Martin et al. (1994) estimated, however, the primary ^{238}Pu source to the Western Scheldt estuary to be the Rupel (CEN Mol) with an annual release (1979-1980) of 3 mCi/yr (9×10^4 Bq/yr) and not the power plants of Doel. This input is, however, small compared to the controlled annual release of the reprocessing plants of La Hague (F; Guegueniat and Le Hir 1981) and Sellafield (UK; BNF 1981-1984) with amounts to 6.5 Ci ΣPu in 1977 and 170 Ci ^{238}Pu in 1979-1983, respectively.

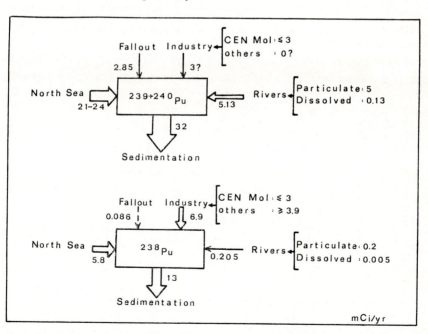

Fig. 7.8. Generalized annual budget of plutonium isotopes of the Western Scheldt estuary between km 50 and 100. Data are in mCi/yr (1 mCi = 3.7×10^{10} Bq).

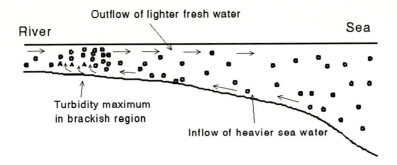

Fig. 7.9. Scheme of particulate matter flows in a tidal estuary.

The Western Scheldt estuary plutonium budget (Fig. 7.8) demonstrates that in a tidally dominated estuary (range of about 4 metres) with large littoral mud flats and a strong turbidity maximum caused by the mixing of particulate matter from both river and marine origin (Fig. 7.9), such a budget must account for multiple sources of input and output. Thus for both $^{239+240}$Pu and ^{238}Pu the inputs from the North Sea (21-24 and 5.8 mCi/yr, respectively) are larger than those of the Scheldt river (5.13 and 0.205 mCi/yr).

The ^{60}Co distribution in sedimentary material of the Western Scheldt estuary exhibits a behaviour different to that of plutonium isotopes (Fig. 7.10), as a prominent peak near 60 km from the sea is evident (nuclear power plants Doel). The concentrations decrease both landward and seaward (Duursma et al. 1985). A decrease of ^{60}Co in the suspended matter has been observed since 1986 (Fig. 7.10), probably due to a limitation of the radionuclide release by the nuclear plants (Martin et al. 1994).

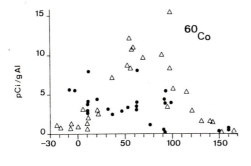

Fig. 7.10. Longitudal distribution of ^{60}Co, normalized to aluminum in recent sediments and suspended matter in the Western Scheldt. Distances are measured upstream from the mouth of the estuary. Δ = 1979-1984, dots = 1986-1988.

[60]Co has a different origin than the plutonium isotopes, which derive from the nuclear fuel. The cobalt isotope is formed by irradiation of metal construction and released due to corrosion in the cooling water system.

7.2 Metals in the Western Mediterranean

Within an interdisciplinary, multi-national programme, entitled EROS 2000 (European River Ocean System), tentative budget studies have been made of various constituents of the Western and North-Western Mediterranean Sea (Fig. 7.11; Martin et al. 1989; Martin and Thomas 1990).

One of the major objectives of EROS 2000 was the determination of the receiving capacity of the sea for riverine and atmospheric contaminants (Martin 1992). Major riverine sources were those of the Rhone (France) and the Ebro (Spain). The Western Mediterranean exchanges sea water with the Atlantic through the Strait of Gibraltar and with the Eastern Mediterranean through the Strait of Sicily (Table 7.2).

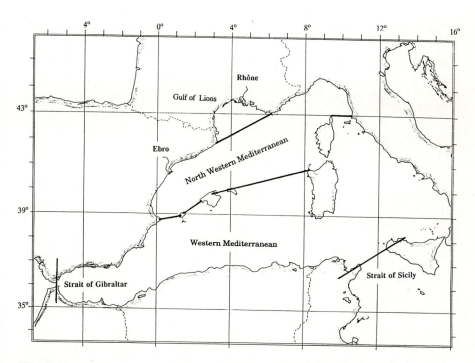

Fig. 7.11. Map of the Western Mediterranean, study region of the EROS 200 programme 1987-1994.

The atmosphere is regarded as a source for contaminants because of the contribution of precipitation, rain and atmospheric dust, predominantly from North Africa. Water circulation in the Western Mediterranean is dominated by exchange of water through the Strait of Gibraltar, where less dense Atlantic surface water enters the Mediterranean over the deeper counter current of denser, higher saline Mediterranean water. The same occurs through the Strait of Sicily, where the eastward surface current flows over the deeper westward counter current.

Table 7.2. Characteristics of the Basins of the Gulf of Lions, the north-western Mediterranean and the western Mediterranean. Data from EROS 2000 discussions, Béthoux (1980) and C. Copin-Montégut (1988).

	Gulf of Lions	North-western Mediterranean	Western Mediterranean
Physical Characteristics			
Surface (x10^3 km^2)	22	280	840
Volume total (x10^3 km^3)	2.2	463	1390
Volume surface (<100 m ")	1.5	28	84
Volume surface (<350 m ")	?	?	294
PM (x10^6 ton)	?	18	55
Fluxes at the Boundaries			
Rain (km^3/yr)			
Evaporation (")	8	100	300
Rivers (")	?	?	1105
Rhone			
Ebro	52		
Total		6.3	
River PM (10^6 ton/yr)	54	86	105
Atmosphere PM (")	5	9	48
Biogenic PM (")	0.5	3	15
Settling PM (")	1		
	40	100	300

Exchange at marine boundaries	*N-w + Western Mediterranean*	*Strait of Gibraltar*	*Strait of Sicily*
Water (x10^3 km^3/yr)	?(in) ?(out)	53.0(in)	38.0(in)
	?(in) ?(out)	50.5(out)	39.8(out)
Suspended matter (x10^6 ton/yr)		26(in)	2.3(in)
		8(out)	14(out)

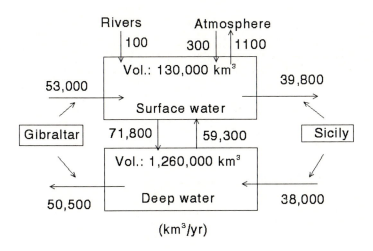

Fig. 7.12A. Budget of water circulation of the Western Mediterranean according to Chou and Wollast (1996) based on data of Béthoux (1980). The volume of surface water is calculated for a layer of 155 m; compare with Table 7.2.

For the Western Mediterranean basin, exchange with the atmosphere is driven by high evaporation rates. Evaporation in this area exceeds precipitation by 805 km^3/yr (Fig. 7.12A) and causes the higher salinity of the Mediterranean water with respect to Atlantic sea water. The annual exchange of surface and deep water is relatively large due to the disappearance of the thermocline in winter resulting in intense vertical mixing.

Since the EROS 2000 inorganic working group has been (1996) unable (Wollast, personal communication) to prepare reliable budgets of trace metals for the Western Mediterranean except for alumiminum (Fig. 12B, Chou and Wollast 1996), the distribution of trace metals between dissolved and particulate phases can only be approached on the basis of earlier EROS 2000 data, published by Tankere et al. (1995) and Yoon et al. (1995).

For this approach also the particulate matter budget of the Western Mediterranean has not yet been agreed upon by EROS 2000 working groups. In particular the exchange between the North-Western and the Western Mediterranean basin has still to be established, because reliable input figures from atmosphere and output by sedimentation are a point of discussion. Nevertheless, some approximation is possible on the stock of particulate matter in this basin. The tentative average concentration of particulate matter can be taken at 39.4 μg PM/l (Fig. 7.13). For a total volume of 1390x10^3 km^3 (Table 7.2), this results in: 5.5x10^7 ton PM in the Western Mediterranean.

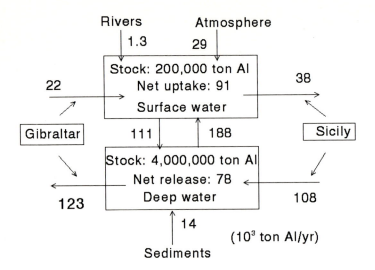

Fig. 12B. Mass balance of dissolved aluminum in the Western Mediterranean. (After Chou and Wollast 1996).

A higher PM concentration was determined by Tankere et al. (1995) with an average PM concentration of 350 μg/l, mainly determined in the Straits of Sicily and Gibraltar.

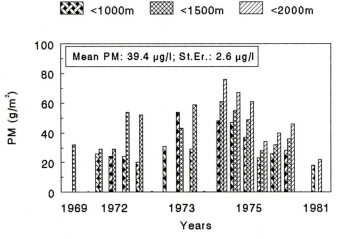

Fig. 7.13. Particulate matter concentrations the Western and North-Western Mediterranean basin (Copin-Montégut, 1988).

As summarized in Fig. 7.14 the average concentrations of dissolved trace metals as determined in Western Mediterranean by the outgoing water of both the Strait of Gibraltar and Sicily (Tankere et al. 1995) agree fairly well with those determined by Yoon et al. (1995) for a wider area of the Western Mediterranean.

The data of Tankere et al. (1995) can further be used to calculate the average distribution coefficients of these trace metals in the Western Mediterranean, by using their particulate trace metal concentration (Fig. 7.15).

From formula 2.18

$$X = \frac{10^8}{S \cdot K_d + 10^6}$$

the percentile distribution or partition coefficient (see also section 8.6) can be calculated.

Supposing that in the open Western Mediterranean the average PM concentration is 39.4 μg/l (Fig. 7.13) and that for the Gibraltar and Sicily street they are 350 μg/l, the % trace metal in solution is then estimated (Table 7.3).

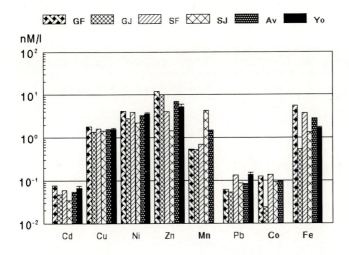

Fig. 7.14. Average dissolved trace metal concentrations of the Western Mediterranean basin, as determined by Tankere et al. (1995), where GF = outgoing water at Gibraltar, February 1992, GJ = outgoing water at Gibraltar, July 1993, SF = outgoing water at Sicily, February 1992, SJ = outgoing water at Sicily, July 1993, Av = Average of these data. Additional average data and their standard error are from Yoon et al. (1995), indicated by the symbol Yo.

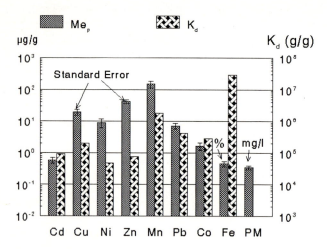

Fig. 7.15. Average particulate trace metal concentrations and their standard errors, as determined by Tankere et al. (1995), and their K_ds (g/g), calculated on the base of the average data (Av) of Fig. 7.14 and the given particulate metal concentrations.

Table 7.3. Tentative percentile dissolved trace metals in the open Western Mediterranean (PM = 0.0394 mg/l) and Sicily and Gibraltar Straits (PM = 0.350 mg/l). In column 3 (Radion.) the K_ds of ^{65}Zn and ^{60}Co are given as determined for Mediterranean sediment (Duursma and Eisma 1973).

		K_d	% dissolved trace metals	
Metals	(x10^6)	Radion. (x10^6)	S = 39.5 μg/l	S = 350 μg/l
Cd	0.093		99.6	96.8
Cu	0.195		99.2	93.6
Ni	0.047		99.8	98.4
Zn	0.076	0.042	99.7	97.4
Mn	1.80	(St.Er. 0.035)	93.4	61.3
Pb	0.408		98.4	87.5
Co	0.285	0.036	98.9	90.9
Fe	28.90	(St.Er. 0.021)	46.8	9.0

Except for iron and partly for manganese, all other trace metals are found to be more then 90 % in solution. This indicates that the major exchange of these trace metals, except manganese and iron, between the Western Mediterranean and the other basins occurs in solution, in spite of their high

K_ds. Thus the balancing of contamination by large rivers, trace metal sources and atmosphere will dominantly occur through exchange of water and not necessarily by settling, although the latter still has to be confirmed.

Exercise 7.1. What is the effective outflow of total cadmium (dissolved and particulate) from the Western Mediterranean to the Atlantic through the Strait of Gibraltar. The average concentrations are (Tankere et al. 1995): Cd-dissolved: 0.078 nmol/l in outflowing (deep) water and 0.027 nmol/l in inflowing (surface) water. The particulate Cd concentrations are: 0.24 μg/g for inflowing (surface) water and 0.020 μg/g for outflowing (deep) water. Use data of Fig. 7.12A and 350 μg/l as PM concentration.

7.3 PCB budgets

7.3.1 Western Mediterranean

PCB flux and budget studies were also performed within the project EROS 2000 by Spanish and IAEA laboratories (Tolosa et al. 1996). The main objective was to determine the behaviour of PCBs in the Western Mediterranean for the atmosphere, sea water and sediment compartments. A tentative mass balance (Fig. 7.16) indicates a range of concentrations of 100-300 pg ΣPCB/l for a volume of Western Mediterranean water of 1.5×10^{15} m^3 (surface 6.3×10^{11} m^2 and average depth of 2380 m).

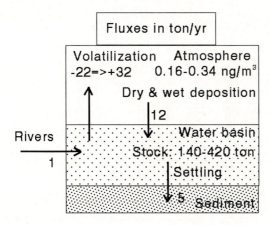

Fig. 7.16. Budget of ΣPCBs of the Western Mediterranean in ton/yr.

According to Tolosa et al. (1996) a significant proportion of the PCB load enters the marine environment through wet and dry deposition, while a counter process of volatilization seems to be similar to observations for other surface waters (Baker and Eisenreich, 1990). Baker and Eisenreich's data suggest that the atmospheric concentrations seem to have been stable for the last decades since measured values (0.16 ng/m^3 in the open sea atmosphere and 0.34 ng/m^3 for coastal regions) are of the same order of magnitude as those in air above the North Pacific (0.012-0.39 ng/m^3) and North Atlantic (0.072-0.60 ng/m^3) during 1989-1990 (Iwata et al. 1993).

With a stock of 140-420 ton ΣPCB in the Western Mediterranean basin, an input of 13 ton/yr by rivers, dry and wet deposition, and an output by settling of 5 ton/yr, this stock may have been build up in 17.5 to 52.5 years, neglecting volatilization, which ranges between -22 (deposition) to +32 (volatilisation) ton/yr. It is difficult to determine a residence time, since the system may not yet be in steady state.

7.3.2 Baltic

In a recent study by Schulz-Bull et al. (1995), 31 dissolved and particulate PCB congeners have been measured in Baltic sea waters between 1988-1991 (Fig. 7.17).

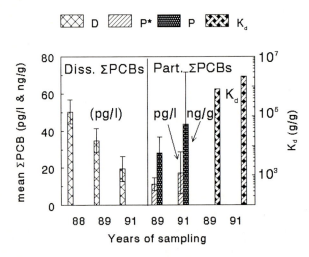

Fig. 7.17. Mean dissolved and particulate ΣPCB concentrations of the Baltic ± 1 standard error D = dissolved, P* = particulate (μg/l), P = particulate (ng/g).

Unfortunately the concentrations of particulate PCB congeners were only given in pg/l instead of pg/g PM (PM = particulate matter) where PM data were lacking. In the latter case it would have been possible to determine the ratio of the dissolved and particulate PCBs (K_d as defined by formula 2.10). However, taking the PM average value of 0.4 ± 0.2 mg/l (0-400 m) as given by Bernard and van Grieken (1989) and the mean particulate PCB values found in pg/l of 11.3 for 1989 and 17.4 for 1991, respectively, this gives estimates of 28.25 ng PCB/g PM in 1989 and 43.5 ng PCB/g PM in 1991. The respective K_ds (g/g) are for 1989 and 1991 0.81×10^6 and 2.2×10^6.

7.3.3 Antarctic atmosphere

The fact that DDT was detected in the Antarctic between 1988-1990 (Larsson et al. 1992) indicates that, in spite of the bans on its present production and use, this substance or its derivates are still detectable in remote regions. These authors found mean ΣDDT (p,p'DDT + p,p'DDE) levels of 3.7 pg/m^3, (range: 0 and 17 pg/m^3 with one exceptional value of 148 pg/m^3, which is not used further for the calculations; Fig. 7.18). Somewhat higher concentrations of Lindane (γHCH) and PCBs were found in air samples.

From the same period, Larsson et al. (1992) have found p,p'DDE and Lindane contents in zooplankton ranging between 19 and 3320 ng/g lipid for Lindane and between 3 and 85 ng/g lipid (fat) for p,p'DDE (with one exceptional value of 354 ng/g).

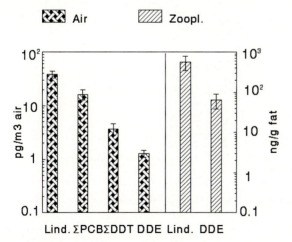

Fig. 7.18. Chlorinated substances Lindane, PCBs, DDT and DDE in air and zooplankton in the Antarctic in 1988-1990 (Larsson et al. 1992). One extreme value was excluded.

Using the average values of Fig. 7.18, The K_ds (lipid/air) can be calculated, in spite of the fact that the contact between the Lindane and DDE in air and zooplankton passes through the water compartment.

$$K_d \ (PCB \ lipid/air) \ = \ \frac{ng \ PCB/g \ lipid}{pg \ PCB/m^3 \ air}$$
$$= \left(1.25x10^9\right)\frac{pg \ PCB/m^3 \ lipid}{pg \ PCB/m^3 \ air}$$

(7.1)

where the specific weight of lipid is taken as 0.8 g/ml (Fig. 7.19).

Keeping all units of concentration in weight/volume (pg/m^3), the correlation between the different K_ds is:

$$K_d \ (PCB \ lipid/air) \ = \ K_d \ (PCB \ lipid/water) \ x \ K_d(PCB \ water/air)$$

(7.2)

These calculated values are given for the Antarctic in Fig. 7.19, supposing the K_d (lipid/water) for lindane is about 10^5 and for DDE 10^6. The result is that the K_ds (water-air) for DDE are much higher (40,600 (m^3/m^3); Fig. 7.19) than determined for ΣDDT in the oceans (59.2 (m^3/m^3); Fig. 7.20), which indicates a non-equilibrium in the Antarctic between atmosphere (as source) and the water compartment.

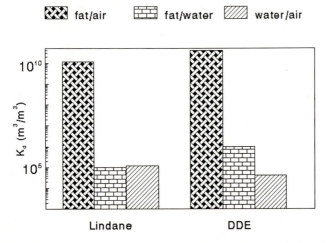

Fig. 7.19. Calculated ranges of K_ds for lindane and p,p' DDE, based on data in Fig. 7.18.

7.3.4 Hypothetical global PCB equilibrium
between atmosphere, water and man

If we bring together the scarce but reliable concentrations of PCBs and DDT (or ΣDDT) in atmospheric, oceanic water and particulate and mussel samples, it is hypothetically possible to calculate a distribution budget between these compartments. The global environment has been exposed to these products since the 1940s, but releases over the past several decades into the environment have been curtailed. Despite unequal east-west (dominant) and north-south (slower) atmospheric circulation patterns, it is at least theoretically possible to assume that some kind of equilibrium has been attained world-wide between the air, oceans and organisms compartments.

(7.20)

Table 7.4. List of sampling stations (Iwata et al. 1993), mentioned in Figs. 7.20A and B.

No.	Sampling location	No.	Sampling location
1	Chukchi Sea (n=3)	10	Red Sea (n=1)
2	Bering Sea (n=4)	11	East China Sea (n=6)
3	Gulf of Alaska (n=3)	12	South China Sea (n=6)
4	N. North Pacific (n=12)	13	Strait of Malacca (n=1)
5	North Pacific (n=8)	14	Celebes Sea (n=1)
6	Caribbean Sea (n=1)	15	Java Sea (n=1)
7	Gulf of Mexico (n=1)	16	Bengal Bay, Arabian Sea (n=7)
8	North Atlantic	17	Eastern Indian Ocean (n=5)
9	Mediterranean (n=2)	18	Southern Ocean (n=5)

Using the ΣDDT and ΣPCB concentrations as found and cited by various scientists, and in particular those of Iwata et al. (1993) (Table 7.4; Figs. 7.20A and B; Table 7.5) it is possible to make a rough estimation of the global abundance of these organochlorines in the lower atmosphere and upper ocean water.

Supposing that both the organochlorines ΣDDT and ΣPCB are mainly distributed in the lower atmosphere below the high cloud level. This is the location of more or less continuous exchange between gaseous and cloud-dissolved states, and a consequent precipitation to the earth-ocean surface. This assumption leads to the conclusion that most of the ΣDDT and ΣPCB will be found in the lower atmosphere up to a height of 10 km above sea level.

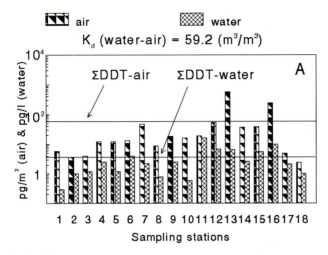

Fig. 7.20A. ΣDDT in atmosphere and oceans of various sampling stations in 1989-1990, as determined by Iwata et al. (1993).

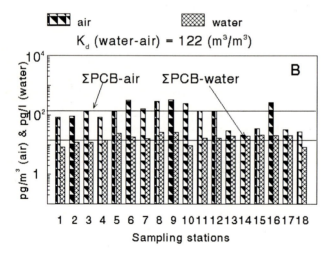

Fig. 7.20B. The same for ΣPCB. The numbers (Table 7.4) represent the sequence of stations as given by the authors.

Table 7.5. Summary of average ΣDDT, PCB and HCH concentrations (in pg/m³) in remote atmosphere (Atm.) and oceans (Oce.) (dissolved).

	ΣDDT		ΣPCB		ΣHCH	
Source	Atm.	Oce.	Atm.	Oce.	Atm.	Oce.
Duursma et al. (1986) (North Sea)				˜0.050		
Villeneuve (1986) (Mediterranean)			˜33			
Baker and Eisenreich (1990)			10-1600			
IMW (1994) (S. & C. America coast)		˜0.84 (calc. from K_d)		˜0.88 (calc. from K_d)		
Larsson et al. (1992) (Antarctic)	3.68		16.2		38.6	
Iwata et al. (1993) (Various oceans)	62.6	0.0037	142.8	0.0174	849	0.558
Schulz-Bull et al. (1995) (Baltic)				˜0.035		
Tolosa et al. (1996) (Mediterranean)				˜0.05 (open sea) ˜0.2 (coast)		

The partial global stock in atmosphere and upper ocean can subsequently been calculated from the atmospheric concentration ratio of ΣDDT or ΣPCB and oxygen at sea level, multiplied by the fraction of total global oxygen in the atmosphere up to 10 km height (Fig. 7.21).

The cumulative reservoir of oxygen to 10 km height/m² is given by the two formulas (7.6) and (7.7):

$$x(height) = -16120 \cdot \log P_x + 48870 \quad (m) \tag{7.3}$$

where P_x is the barometric pressure in mBar at height x, and:

$$O_{cum} = 9.35 \int_{x=0}^{x=10,000} \frac{P_x}{P_0} dx \quad mol/m^2 \tag{7.4}$$

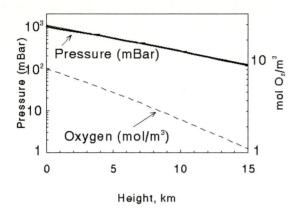

Fig. 7.21. Atmospheric pressure and oxygen content as function of height (Duursma and Boisson 1994).

where P_0 = 1000 mBar.

The solution of equations 7.6 and 7.7 gives an oxygen amount of 5.35×10^4 mol O_2/m^2, which is 76% of the total oxygen present up to $x = \infty$. Using values of atmospheric ΣDDT and ΣPCB concentrations at sea-level (Iwata et al. 1993; Fig. 7.20A and B) of: 62 pg/m^3 and 142 pg/m^3, respectively and the ratio [Global oxygen content in mol until 10 km height]/[mol O_2/m^3 at sea level], which is $0.76 \times 3.75 \times 10^{19}/9.35 = 3.05 \times 10^{18}$, the global atmospheric stocks become 189 ton for ΣDDT and 433 ton for ΣPCB.

Again using data from Iwata et al. (1993) (Fig. 7.20A and B) for the mean dissolved ΣDDT and ΣPCB concentration in the oceans of 3.7 pg/l and 17 pg/l respectively, and considering these concentrations to be valid for the upper 400 m of the ocean ($400 \times 3.6 \times 10^{14}$ m^3), the stock can be calculated as 533 ton ΣDDT and 2448 ton ΣPCB (Fig. 7.22).

How are man and aquatic organisms in equilibrium with these compartments? The hypothesis is made that DDT in man, through the supposed atmosphere-blood-body fat pathway, is in equilibrium with atmospheric DDT. In mussels the pathway is atmosphere-water-body fluid and body fat. The strikingly similar concentrations of ΣDDT in both body fat of human populations from different continents (Fig. 5.19) and of mussels (Fig. 7.23) around South and Central America (Fig. 5.9A), seems to strongly suggest an apparent equilibrium existing through the atmosphere.

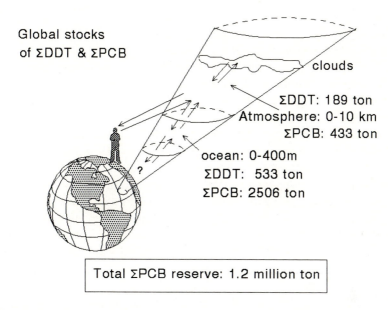

Fig. 7.22. Tentative global stocks as estimated for ΣDDT and ΣPCB.

The same conclusion is less easily reached for PCBs, since their distribution has larger peaks in industrialized regions. Nevertheless the ΣPCB levels in mother's milk for the Netherlands and Germany were between 1000 and 2400 μg/kg (Fig. 5.20), while those found for the South & Central American mussels were about 878 μg/kg.

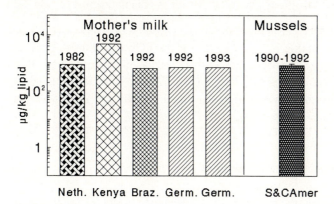

Fig. 7.23. Average ΣDDT concentrations in mother's milk an mussels; from Figs. 5.8A and 5.20.

These estimates are still of the same order of magnitude, and a similar world-wide contamination of man by atmospheric distribution of PCBs has to be expected at some point in the future.

It is not in the context of this book to elaborate on the toxicity of the organochlorines in man at the levels observed. One can only state that the world-wide danger of contamination by DDT is not past. The toxicity of individual PCB-congeners differs, while changes in composition also take place (Boon et al. 1992). Furthermore, it is not clear whether ΣDDT contamination is really diminishing; the figures for mother's milk since 1970 do not support such an indication. Iwata et al. (1993) cites an annual use in India of 20,000 ton/yr, while DDT spraying is still applied in China. However, it is reasonable to presume that, unless production and use of DDT does not increase, future ΣDDT contamination will probably not proceed further than presently observed.

The scenario might be different for PCBs. The world waste stock of PCBs (1.2×10^6 ton) is several orders of magnitude higher (Fig. 5.3) than estimated for the lower atmosphere (433 ton) and the upper 400 m of the oceans (2448 ton) (Fig. 7.22). A release of even 0.1% of the world waste stock would increase atmosphere and ocean concentrations by a factor of 40. There will be no limit for evaporation to the atmosphere and dissolution in the oceans (Fig. 5.3). Consequently living organisms, either aquatic or terrestrial would be contaminated equally and proportionally.

It is certainly worthwhile to eradicate the PCB wastes with an efficiency above 99.99%, otherwise these contaminants remain a kind of environmental time-bomb. Nevertheless, over long time scales, bottom sediments can potentially accumulate a great part of PCBs, as is calculated in exercise 7.2.

Exercise 7.2. When taking $K_d = 10^6$ (g/g) for the PCB distribution between bottom sediments and water (Fig. 7.17) and a mean dissolved PCB concentration in the oceans of 17 pg/l (Fig. 7.20B), what is the stock of PCBs in a 5 cm bottom layer, supposing equal distribution in this bottom layer? Ocean surface is 3.6×10^{14} m^2.

7.4 The food chain accumulation paradox

In the preface the remark was given *(...) hypotheses, which differ principally from the often dogmatized theory on accumulation of contaminants in food chains, (...)*. This remark requires a discussion. The dogma on accumulation in the food chain is so wide-spread that it even appears in education books and cartoons for children.

It certainly cannot be disputed that contaminated food once ingested will contaminate the consumer (Dewailly et al 1994; Ayotte et al. 1995), where the level of contamination may reach high values when this process is chronic (Harding et al in press). The models as presented in Figs. 2.15 and 5.21 would be applicable for determining the uptake of contaminants through ingestion by fish or man, respectively. Indeed feeding habits determine the level of contamination, as was shown in Fig. 5.24 with respect to PCBs and DDT in man. Although the hypothesis of accumulation in the food chain is based on data and modelling, the major question is usually not answered: to which extent will contamination of the consumer proceed and which equilibrium between uptake and loss may be attained once a steady state situation is reached between levels of contamination in the environment.

What has been demonstrated for aquatic gill-breathing organisms in chapter 5, are that PCB equilibria may be reached between the organism and water compartment, suggesting a kind of chemical solubility ratio between PCBs in lipids and those in water. Partition coefficients were determined of the order of 10^5-10^6 [(μg/g lipid):(μg/g water)], to be independent of the kind of organism and their place in the food chain. This fact is taken as a leading principle for explaining equilibria between uptake and loss for aquatic organisms. It was further demonstrated in section 5.3 for fasting eel, that the loss process requires an active metabolism, since otherwise the release of PCBs from lipids to blood and water compartments was suppressed.

For non-aquatic organisms such an equilibrium between uptake and the water compartment cannot exist, and loss can only occur through excretion and respiration. Since these processes may be less effective than the contact between water and blood through gills, mammals or birds may suffer from chronic uptake of PCBs or DDT and its metabolites. This finding was originally the basis for the hypothesis of accumulation in the food chain.

Nevertheless, uptake by and loss of organochlorines from mammals and birds might be approached in an identical manner as for aquatic organisms. Only the water compartment should be replaced by the atmosphere compartment. When taking additionally a time factor into consideration, which can be expressed in a biological half-time ($t_{b\frac{1}{2}}$), the difference between mammals, birds and aquatic organisms is less distinct. The $t_{b\frac{1}{2}}$ is probably larger for mammals and birds than for aquatic organisms, at least for organochlorine contaminants. Those for mammals and birds are little studied, since the hypothesis of uptake and loss of PCBs and DDTs through respiration has been little fostered.

The data, as presented in section 5.5 and 7.3.4, indicate that such a mechanism must exist in order to explain the global ΣDDT levels in man, which match those in mussels around South America. A global equilibrium exists between ΣDDT in the atmosphere, water and organisms (man and mussel) compartments.

The dogma of accumulation in the food chain should therefore be replaced by an equilibria-seeking process, which explains both accumulation of contaminants and their chemically derived tendencies to attain equilibria between the major environmental compartments.

8 Nuclear waste in the Kara Sea

8.1 Introduction

Risk assessments provide the foundation of a scientifically sound and defensible method of evaluating the impacts of contaminants on both the biotic and abiotic components of an environment. The procedure involves identifying a hazard, examining the relationship between a contamination event and determining its effects on humans and the environment. Field studies, laboratory tests and models are all used to quantify the magnitudes and probabilities of effects. The information is then used to make decisions on how to balance acceptable risks against the costs of risk reduction (Paustenbach 1989).

This chapter presents a summary of the procedures used to evaluate specific risks to the environment and human health associated with solid and liquid nuclear wastes that reside in the Kara Sea, a shallow marginal sea located adjacent to the former Soviet Union. The major focus is on methods to estimate distribution coefficients of radionuclide contaminants, with discussion on uncertainties and limitations of the various methods. This information is presented in the context of the goals of the impact assessment for the Kara Sea. These goals are to determine nuclear waste inventories, estimate present-day levels of contamination, evaluate the potential for future radioactive releases, assess current and possible future impacts on human health and design a long-term plan for the wastes.

The work presented on radioactive waste in the Kara Sea encompasses the efforts of many individuals working on aspects of the problem primarily under the auspices of the International Atomic Energy Agency (IAEA) International Arctic Seas Assessment Program (IASAP). IASAP is coordinated by K-L Sjoeblom and others within the IAEA organization in Vienna. Most of the activities discussed are based on research originating in the IAEA Marine Environment Laboratory, Monaco. The work has particularly benefited from cooperations with the Joint Russian-Norwegian Expert group for Investigations of Radioactive Contamination in the Northern Areas, the coordinators and investigators of the US Office of Naval Research Arctic Nuclear Waste Assessment Program and the Norwegian Defense Research Establishment.

8.2 History of radioactive waste dumping in the Kara Sea

As a consequence of nuclear development activities, many chemical and radioactive contaminants have accumulated in the global marine environment during the last half century.

The Arctic in particular has been impacted by several well-known sources of radioactive contamination: nuclear weapons testing, releases from nuclear installations, European reprocessing plants and the Chernobyl accident (Table 8.1). In 1992, Russian authorities revealed a new source of radioactive contamination to several shallow Arctic seas (White Book 3 1993). Beginning in the mid-sixties, substantial quantities of solid and liquid nuclear wastes were discharged into shallow bays located along the margin of the island of Novaya Zemlya (70-80° N, 50-70° E) and into the Novaya Zemlya Trough (Fig. 8.1).

The items dumped included six nuclear reactor compartments from submarines and a shielding assembly from an icebreaker reactor containing spent fuel, ten reactors without fuel and several thousand containers with low- and intermediate level solid and liquid wastes (Sjoeblom and Linsley 1995). The total activity of the wastes is estimated at 4.5 PBq (4.5×10^{15} Bq) (Sivintsev et al. 1994; Yefimov 1994). As a result of these discharges and dumping activities, a wide variety of radionuclides (i.e. fission products, activation products, and actinides) now reside in the shallow Arctic marine environment.

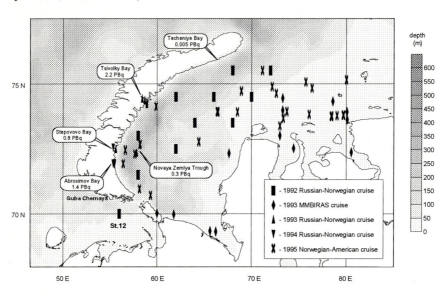

Fig. 8.1. Sampling stations in the Kara Sea 1992-1995. Estimated inventories of radionuclides refer to 1994 and pertain to naval reactors disposed at major dump sites based on IASAP working documents (courtesy of I. Osvath).

Table 8.1. 1993 Inventory Estimates for Radioactivity in Arctic Seas (1PBq = 10^{15}Bq; 1MC = 37 PBq) from Aarkrog (1994).

	^{90}Sr	^{137}Cs
Global fallout	2.6	4.1
Sellafield discharges	1-2	10-15
USSR river discharges	1-5	1-5
Run-off of global fallout	1.5	0.5
Chernobyl	0	1-5
TOTAL	6-11	17-30

The primary list of radionuclides includes radioisotopes of caesium (^{134}Cs, ^{137}Cs), plutonium (^{238}Pu, ^{239}Pu, ^{240}Pu), americium (^{241}Am), strontium (^{90}Sr), cobalt (^{60}Co), nickel (^{63}Ni) and europium (^{152}Eu, ^{154}Eu, ^{155}Eu) (Sivintsev et al. 1994; Yefimov 1994).

In response to public concerns over the present and future impact of radioactive wastes dumped in the Kara Sea, scientists began collecting environmental and contamination data in the Arctic as part of a wide range of national and international monitoring and research programs. In 1993, the International Arctic Seas Assessment Project was undertaken by the International Atomic Energy Agency in collaboration with Member States to assess the risks posed by radioactive contamination to biological and human health and to consider the feasibility of possible remedial actions.

8.3 Inventory of dumped objects

Knowledge of the amount and composition of the radioactive waste, the integrity of containers and the characteristics of the fuel in the different types of dumped reactors is a critical first step in assessing the risk of contamination expected from the release of radionuclides from the dumped objects (Table 8.2). Such an evaluation of the dumped objects was conducted by experts familiar with the former Soviet Union nuclear development programs (Sjoeblom and Linsley 1995). Reactors containing spent nuclear fuel pose the highest risk of contamination to humans and the environment if leakage should occur. The total activity of the reactors at the time of dumping was estimated at 37 PBq (Sjoeblom and Linsley 1995). This is less than the original estimate of 89 (Table 8.2) provided in May 1993 by the Russian Federation. Most of the reactors met with an accident after a very short period of operation which was not taken into consideration in the initial inventory estimates. Due to radioactive decay since the dumping occurred the total activity has been reduced to 4.7 PBq. Most of the reactors are shielded with metal or concrete and filled with a polymer, furfural. Designers estimate that reactors filled with furfural are safe for several hundred years.

Table 8.2. Data on nuclear reactors dumped near Novaya Zemlya (from Sjoeblom and Linsley 1995).

	Year of dumping	Depth[a] of dumping (m)	Dumping Unit	No. of Reactors Without spent nuclear fuel	With spent nuclear fuel	Total activity (PBq = 10^{15}Bq) At time of dumping	1993/94
Abrosimov Bay V = 3.8x10^8 m^3	1965	10-15[b]	Reactor Comp.[c]	1	1	29.6	0.655
		10-15[b]	Reactor Comp.[c]	-	2	14.8	0.727
		20	Reactor Comp.[c]	2	-	d	0.009
	1966	20	Reactor Comp.[c]	2	-	d	0.005
Tsivolka Bay	1967	50	Reactor Comp.[c] and box of fuel	3	0.6	3.7[d]	2.2
East of Novaya Zemlya Trough	1972	300	Reactor	-	1	29.6	0.293
Stepovogo Bay V = 2.0x10^8 m^3	1981	30[b]	Submarine	-	2	7.4	0.838
Techeniye Bay	1988	35-40	Reactors	2	-	d 89	0.005
TOTAL				10	6.6		4.7

[a] The data on depths of dumping were provided in May 1993 by the Russian Federation;
[b] Data obtained during joint Norwegian-Russian scientific cruises in 1993 and 1994.
[c] Comp. = compartment
[d] Reactors without spent nuclear fuel, not more than 3.7 PBq total.

8.4 Environmental concentrations of radionuclides

Data on current and historic levels of radioactive contamination in Arctic seas is necessary to determine the extent to which leakage of radioactive waste has already occurred. These data have been gathered during numerous research expeditions to the Arctic during the past four years sponsored by a variety of national and international agencies. These expeditions have resulted in an extensive array of information on radionuclide levels in water, sediment and biota samples for a region of the world ocean where previous information was sparse. The environmental data are archived in several databases including the IAEA Marine Environment Laboratory Global Marine Radioactivity Database (GLOMARD) (Hamilton et al. 1994; Povinec et al. 1995). The present-day levels of radionuclides in the Kara Sea are not appreciably elevated above the levels expected from nuclear weapons test fallout and land-based sources (Table 8.3).

Table 8.3. Sediment and water (surface and bottom) activities for selected radionuclides from stations in the open Kara Sea (compiled from Hamilton et al. 1994 and Strand et al. 1994). Only stations with measurable activities were used in the determinations.

	Median	Maximum	Minimum	
WATER				
^{137}Cs (Bq/m^3)	7.8	20.4	3.3	n = 39
^{238}Pu (mBq/m^3)	0.3	1.4	0.1	n = 17; n.d. = 18
^{239,240}Pu (mBq/m^3)	6.0	16.0	1.8	n = 35; n.d. = 1
^{90}Sr (Bq/m^3)	4.0	12.1	3.0	n = 39; n.d. = 13
^{241}Am (mBq/m^3)	0.85	0.2	2.6	n = 16
SEDIMENT				
^{137}Cs (Bq/kg)	24.5	32.0	17.9	n = 8
^{239,240}Pu (Bq/kg)	0.78	1.25	0.44	n = 8
^{238}Pu (Bq/kg)	0.02	0.05	0.01	n = 8
^{241}Am (Bq/kg)	0.28	0.46	0.20	n = 6

n.d. = non detectable; n = number of measurements

Inventories of ^{137}Cs, one of the primary waste-related radionuclides, increase uniformly with water depth in sea water and decrease uniformly in sediment suggesting that no additional sources of ^{137}Cs exists in the Kara Sea (Fig. 8.2) (Hamilton et al. 1994). One exception was noted at Station 12 which exhibited a higher than expected ^{137}Cs inventory in sediment. Station 12 is located in the Barents Sea near the entrance to the Kara Sea and probably reflects the close proximity of the station to the Guba Chernaya underwater nuclear test site.

The highest sediment contamination appears to be associated with leakage from dumped containers. Near the containers, concentrations of some waste-related radionuclides are slightly elevated in near-surface sediments (Osvath et al. 1995). An investigation of sediments from one of the dumpsites in Stepovogo Bay shows relatively high concentrations of ^{137}Cs (30-290 Bq g^{-1}) and trace amounts of ^{60}Co.

Based on the age-to-depth relationship at this location, leakage occurred in the early 1980s (Fig. 8.3). The age-to-depth relationship was determined from the sediment profile of ^{210}Pb as interpreted by the Sediment Isotope Tomography (SIT) method (Carroll et al. 1995; Liu et al. 1991).

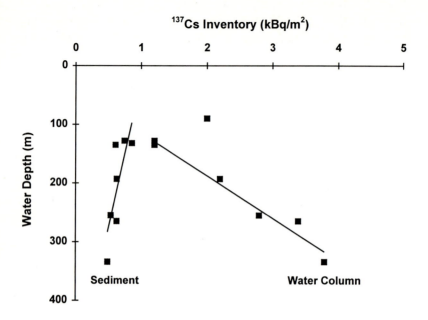

Fig. 8.2. [137]Cs inventories in sediment and seawater for the Kara and Barents Seas as a function of water depth (from Hamilton et al. 1994).

8.5 Predicting future radionuclide releases

While it is clear from the environmental surveys that most of the waste in the Kara Sea is still effectively contained, there are concerns about the long term integrity of waste containers. In order to identify potential risks in the future, estimates are needed of release rates from dumped objects and where and how the radionuclides will be dispersed if mobilized. From studies of the weakest points of the protective barriers, estimates of release rates for various radionuclides have been made (Sjoeblom and Linsley 1995). Releases are based on two possible scenarios: (i) a continuous release, (ii) an instantaneous release and (iii) a combination of (i) and (ii).

Since 1965 several reactor compartments containing spent nuclear fuel have resided in Ambrosimov Bay. For a postulated release scenario (Fig. 8.4), [239]Pu releases begin in 2015. For the next several hundred years a gradual release of radioactivity occurs as corrosive forces breakdown the waste containers. This is followed by a major failure of containment structures resulting in a total release of the remaining radioactivity as some of the last containment barriers collapse.

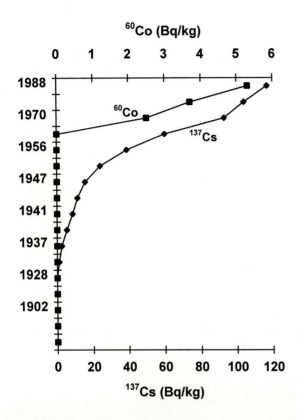

Fig. 8.3. Input histories of ^{60}Co and ^{137}Cs to bottom sediments near a dump site in Abrosimov Bay.

8.6 Modelling radionuclide dispersion

Waste release scenarios such as the one shown in Fig. 8.4 are used as input parameters to computer models designed to predict concentrations and transport pathways of radionuclides after mobilization. Modelling exercises within the IAEA program, IASAP, are being conducted at laboratories in Denmark, Japan, the Netherlands, Russia, Switzerland, the United Kingdom, and the IAEA-MEL (Sjoeblom and Linsley, 1995; Hamilton et al. 1994). Other modelling and assessment programs exist in Norway (NDRE 1995) and the United States (ONR 1995).

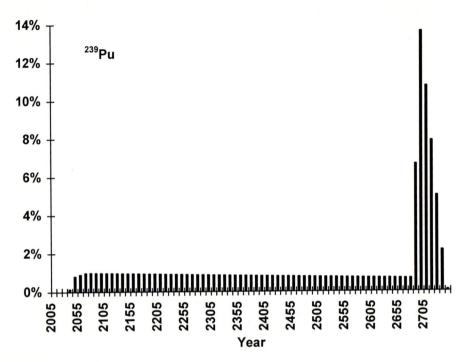

Fig. 8.4. Hypothetical release scenario for ^{239}Pu released in Abrosimov Bay.

Obtaining basic information on the potential uptake of contaminants by marine particles and seawater for the Kara Sea is fundamental to an evaluation of the transport pathways of radionuclides. Numerical models depend on a parameter, known as the partition coefficient, to establish the equilibrium between PM and seawater radionuclide concentrations in their models.

A partition coefficient ranges from zero to one and is a function of K_d, in m^3/kg and suspended sediment concentration S, in kg/m^3, where sediment concentration is operationally defined as the dry weight of material collected on a 0.45 μm pore size filter. The partition coefficient for seawater (PC_w) is defined in radionuclide dispersion models as,

$$PC_W = \frac{1}{1 + K_d \cdot S} \tag{8.1}$$

PC_w represents the fraction of the total radionuclide activity in a parcel of water that remains dissolved (Fig. 8.5; see also section 2.3.2.2, equation 2.28). For this discussion of the modelling program, a summary is given of the difficulties encountered in establishing partition coefficient values from the governing variables, K_d and S, followed by the approach used to determine partition coefficient ranges and uncertainties for use in the waste assessment models. Finally, the types of models used in the waste assessment are discussed and examples of modelling exercises are presented.

> **Exercise 8.1.** PC_S, the partition coefficient for sediment, represents the fraction of the total radionuclide activity that sorbs onto particulate matter in a parcel of water. Write the equation in terms of K_d and S that describes PC_S.

8.6.1 Radionuclide exchange processes

The uptake potential for radionuclides on PM is a function of the relative influence of the chemical properties of the radioisotopes, sediment characteristics and the moderating influence of the oceanic environment.

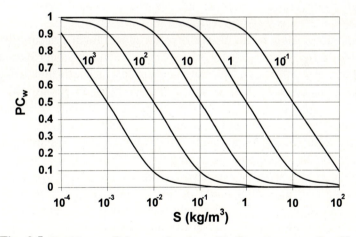

Fig. 8.5. Partition coefficients as functions of sediment concentration (S) and distribution coefficients (K_ds).

Many possible exchange reactions may occur at the sediment—solute interface. Reaction types as well as specific characteristics of particulate matter and water quality are summarized in Table 8.4. The interdependencies among the variables shown in Table 8.4 help to explain why K_d values determined for individual elements, both in field and laboratory experiments, often vary over several orders of magnitude.

Table 8.4. Reaction types and important characteristics of particulate matter and water which influence K_d determinations.

Exchange reactions	Particulate matter characteristics	Water quality characteristics
Ion exchange	Colloids	pH
Colloidal chemistry	Particulate OM	Dissolved OM
Particle aggregation/dis	Humic coatings	Salinity
Complexation	Clay mineralogy	Redox conditions
Precipitation	Grain size	
Diffusion into pore spaces		
Hydrolysis		

It should be clear from Chapters 2 and 3 that the definition of a distribution coefficient is governed by the choice of models used to describe the exchange processes that occur between particle and seawater compartments. For example, equation 2.10 is used to calculate the equilibrium distribution coefficient. Equation 2.10 is based on the simplest model for ion adsorption to hydrous oxide surfaces, known as the surface complexation model (Stumm et al. 1970; Schindler 1975). To calculate a distribution coefficient using equation 2.10, it is assumed that i) ion exchange between mono-valent chemical species is the primary reaction mechanism, ii) surface reactions quickly reach an equilibrium state and iii) reversible exchange between particle surfaces and seawater occurs. When applied to natural particle assemblages found in the ocean, the additional assumption is made that heterogeneous particle assemblages may be treated as a single surface, thereby obscuring the relative contribution of each component of the assemblage (Balistrieri and Murray 1983, 1984, 1986). The reader is referred to Chapters 2 and 3 for a review of the assumptions and methods of this approach.

Using the surface complexation model, Li (1981) compared partition coefficients determined in short term laboratory experiments with their respective natural partition coefficients in the ocean and showed that most elements are linearly correlated (Fig. 3.14A). This suggests that laboratory experiments reproduced similar partitioning processes that operate under natural conditions in the ocean (Li et al. 1984a).

In oceanic systems where pH fluctuates narrowly around 8, the presence of surface oxides of Mn, Fe, Si, and Al on particles is thought to dominate the rapid exchange process (Li 1981; Balistrieri and Murray 1984). In addition, there is ample evidence that colloids, including organic coatings, play an significant yet poorly-understood role in adsorption (Balistrieri and Murray 1981; Davis 1984; Morel and Gschwend 1987; Dai and Martin 1995). Researchers have performed a variety of laboratory and field investigations on ion exchange to elucidate the governing processes with respect to the surface complexation model. The approach is adopted throughout the remainder of this chapter.

8.6.2 Approaches to determining site-specific K_ds

The paucity of data on the behaviour of radionuclides in shallow Arctic Seas led to studies investigating PM/water distribution coefficients in the Kara Sea. A three-phase approach was developed for the investigation (Fig. 8.6). Theoretical investigation of the sensitivity of partition coefficients to ranges and uncertainties for sediment concentrations and K_ds establishes radionuclides for which more expensive and time-consuming field and laboratory investigations were warranted.

Laboratory investigations provide insight into the influence of specific environmental parameters such as temperature, salinity, sediment characteristics and concentration on K_ds under carefully controlled conditions. Field investigations integrate the influences of all environmental parameters at once and have the advantage of providing information from many different locations in the Kara Sea.

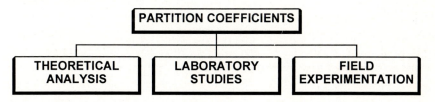

Fig. 8.6. Three-phase approach to analyzing partition coefficients in the Kara Sea.

8.6.3 Theoretical investigation of site-specific K_ds

The actual values for K_d and S used in the calculation of partition coefficients (equation 8.1) are poorly known in most areas of the ocean, and vary spatially and temporally anyway. For the Russian Arctic, data is particularly sparse because access has been limited until recently. Sediment concentrations are expected to vary from a few milligrams per liter of seawater in the open Kara Sea to as high as hundreds of milligrams per liter near active areas on the seabed and near the mixing zones of the Ob and Yenisey Rivers. With little field data available, modellers have relied on K_d values published in the IAEA technical report, 'Sediment K_ds and Concentration Factors for Radionuclides in the Marine Environment' (IAEA 1985; Fig. 3.8). Distribution coefficients in IAEA (1985) were estimated from both stable element geochemical data and the proportions of the particulate phase abundance of the elements that are likely to be exchangeable with the aqueous phase, (also see section 3.2.1.1.5) (Table 8.5). The approach thus provides generic K_d values that would apply to any ocean basin but that are not precise enough to apply in a site-specific assessment of risks such as for the Kara Sea.

Table 8.5. Radionuclide distribution coefficients K_d (m^3/kg) from IAEA (1985) for radionuclides in the Kara Sea.

Nuclide	Mean	Minimum	Maximum
Caesium	3	0.1	20
Plutonium	100	10	1000
Americium	2000	200	20,000
Strontium	1	0.1	5
Cobalt	200	20	1000
Nickel	100	2	500
Europium	500	100	2000

The necessity of having well-constrained values for K_d and S is seen in Figs. 2.13A,B showing the dependency of PC_w for different estimates of K_d over a range of sediment concentrations. In general for low K_d radionuclides in low sediment concentration environments, PC_w will not differ much from unity. Conversely, for high K_d radionuclides in high sediment concentration environments, PC_w will not vary much from zero. Barring these limiting cases, the choice of K_d and S will strongly influence the outcome of modelling experiments performed to assess the transport of radionuclides in the Kara Sea.

For the Kara Sea, transport modellers must choose values for K_d and S to calculate PC_w for each of the various radionuclides being investigated. Rather than simply calculating PC_w based on the average value of K_d and S, a better approach would be to evaluate the influence of many different combinations of K_d and S on PC_w.

The technique used to quantitatively evaluate the impact of uncertainties in K_d and S on PC_w is known as risk analysis. A risk analysis is any technique used to assess the results of making a decision using imprecise or incomplete information. The approach involves developing a model to analyze partition coefficients (equation 8.1). Next, probability distribution functions must be developed for each variable for which a precise value is unknown, i.e. K_d and S. The model is analyzed and a decision is then made based on the outcome.

One way to conduct a risk analysis is by computer simulation. A large number of combinations of values for K_d and S are selected from their probability distribution functions. Those values of K_d and S with higher probabilities will be selected more often than values with low probabilities. The combinations of K_d and S are then used to calculate PC_w. The resulting PC_w values plotted as a function of frequency of occurrence represent the probability distribution function for $PC_w = D(PC_w)$. Simulation thus provides a means of quantifying the uncertainty of PC_w for each of the radionuclides residing in the Kara Sea. As a first step, it provides a framework for designing laboratory and field studies to reduce the uncertainty in PC_w estimates.

For this investigation, IAEA (1985) estimates for mean, minimum and maximum K_d values were used to define the probability distributions $PK_d = D(K_d)$ for K_d and S. The shape of the probability distribution for (equation 8.2) was defined based on an analysis of laboratory data and a few field measurements (Fig. 8.7). The probability distribution function for $S = D(S)$ (equation 8.3) was constructed based on the expected range of sediment concentrations (Fig. 8.7) in the Kara Sea.

The probability distribution function $D(K_d)$ shown for caesium is the exponential function:

$$D(K_d) = \frac{1}{K} e^{\left(-\frac{K_d}{K}\right)}; \ 0 \le D(K_d) \le 1, \tag{8.2}$$

where K is the average value of K_d and K=3 (from Table 8.5). The probability distribution function for S is of the form:

$$D(S) = \frac{S}{S_*^2} e^{\left(-\frac{S}{S_*}\right)}; \quad 0 \le D(S) \le 1,$$ (8.3)

where S_* is the average value of S is equal to 0.005 kg/m³ = 5 mg/l in the open Kara Sea (Fig. 8.7).

A simulation is then performed using these two probability functions to calculate PC_s or PC_w for each radionuclide. The expected range of PC_w for each radionuclide is then given (Fig. 8.8) as the cumulative probability values P(10%), P(68%) and P(90%).

The simulation analysis shows that transport modellers must accept small errors in PC_w only for caesium and strontium. All other radionuclides exhibit large variations around the most probable value P(68%) value suggesting a serious problem in accurately defining partition coefficients for these radionuclides based solely on the IAEA (1985) estimates.

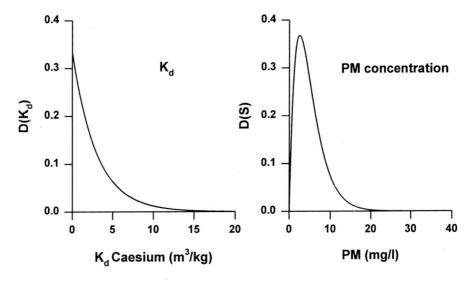

Fig. 8.7. Probability distributions for radionuclide distribution coefficients and sediment concentration used in PC_w simulations. Example K_d distribution is for caesium.

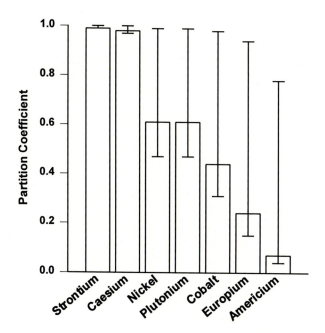

Fig. 8.8. Probabilities (P(10%), P(68%) and P(90%)) for water partition coefficients determined by computer simulation for radionuclides of concern in the Kara Sea.

Alternative to simulation is an analytical approach. To conduct a risk analysis using an analytical approach, the probability distribution function $D(PC_w)$ is defined mathematically from the probability distribution functions of each of the independent variables, K_d (equation 8.2) and S (equation 8.3). Once the probability function has been described as a mathematical formula, the value of PC_w associated with the probability of obtaining that value can be solved explicitly from the equation. The cumulative probability function $D(PC_w)$ represents the probability of obtaining a value of less than or equal to PC_w. Carroll and Lerche (1996) show that, for partition coefficients, the cumulative probability function (for water) is closely approximated by:

$$D(PC_W) = 1 - \left[\frac{D^2 K_* S_*}{0.36} \left(1 + \frac{D^2 K_* S_*}{0.36} \right)^{-1} \right] ; \quad 0 \le D(PC_W) \le 1, \tag{8.4}$$

The advantage of the analytical approach over simulation is that if the form of the cumulative probability function is explicitly derived as for equation 8.4, computer simulations for individual elements are not necessary as long as they satisfy equation 8.2. This is because the probabilities can be calculated directly simply by knowing the average values of K_d and S. The behaviour of $D(PC_w)$ is then generalized for all K_d and S for the probability functions given in equations 8.2 and 8.3

8.6.4 Laboratory determinations of distribution coefficients

As a follow-on to theoretical analyses, laboratory experiments were conducted to determine which environmental parameters principally influence K_ds in the Kara Sea (Carroll et al. 1996). Conducting laboratory experiments with sediment and seawater from a particular area provides only a qualitative understanding of the influence of specific variables at a specific location. Oftentimes, however K_d determinations may be achieved only through laboratory experimentation because of the logistical and regulatory difficulty associated with conducting in situ experiments with radioactivity on an oceanographic research vessel.

The influence of adsorption kinetics and changes in the ionic strength of seawater and sediment concentration were evaluated for surficial bottom sediments collected from Abrosimov and Stepovogo Bays where radioactive debris was located during a 1994 joint Russian-Norwegian expedition. Abrosimov and Stepovogo Bays are shallow, fjord-type estuaries with average water depths of 10-20 m; Stepovogo Bay contains a small central basin where water depths reach 50 m (Fig. 8.9). The sediment samples taken from the bays and used in the experiments were >94% (dry weight) in the <63 μm size fraction (Fig. 8.10).

Laboratory experiments were conducted using the radiotracers, ^{241}Am ($t_{1/2}$=433y), ^{109}Cd ($t_{1/2}$=462d), ^{60}Co ($t_{1/2}$=5.3y), ^{134}Cs ($t_{1/2}$=2.1y), ^{152}Eu ($t_{1/2}$=8.6y), ^{106}Ru ($t_{1/2}$=372.6d) and ^{85}Sr ($t_{1/2}$=64.9 d), and correspond to key radionuclides (or their analogues) that are listed in radionuclide inventories for the Arctic seas (Mount et al. 1993; Sivintsev et al. 1994; Yefimov 1994). An organically bound form of radio-cobalt (^{57}Co-cobalamine, $t_{1/2}$=272 d) was also included as an analogue for ^{60}Co which might be present in nuclear waste containing organic complexing agents. In addition, the behaviour of a few trace metals ^{110m}Ag ($t_{1/2}$=249.8d), ^{133}Ba ($t_{1/2}$=10.8 y), and ^{54}Mn ($t_{1/2}$=313 d), was investigated.

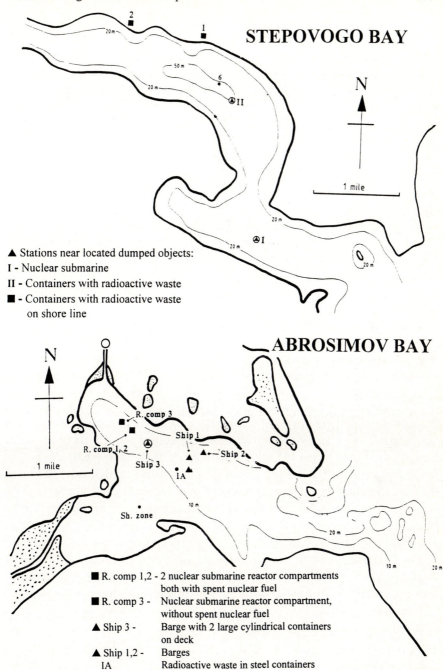

Fig. 8.9. Sampling stations and location of dumped objects in Stepovogo and Abrosimov Bays (Cruise Report 1994). Sediment samples were retrieved at station 6 in Stepovogo Bay and station 1A in Abrosimov Bay.

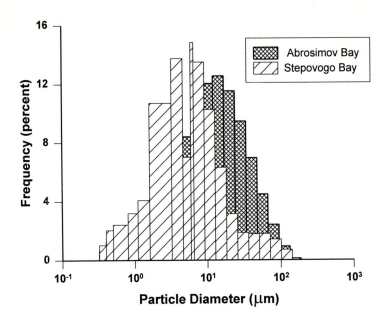

Fig. 8.10. Comparison of grain size distributions for sediments used in laboratory adsorption experiments (Carroll et al. 1996).

The experimental design was similar to the methodology presented in section 3.1.1. A known concentration of sediment was added to bottles of filtered Kara Sea water. The tracers were then added to the bottles and the pH of the water was adjusted back to the pH value for Kara Sea water (pH = 8.0). All bottles were maintained in the dark at 1.5 ± 0.5 °C and gently swirled twice each day. At each sampling date, 20 ml subsamples were filtered through individual 0.2 μm 47 mm Nuclepore™ filters. Filters were rinsed with 2 ml of 0.22 μm filtered seawater at the same salinity as the subsamples. A control bottle containing no sediment was prepared to (i) detect the presence of foreign particles in the flasks, (ii) monitor adsorption onto bottle walls and filters and (iii) identify experimental artifacts such as precipitation of the radionuclides from solution. The radioactivities in filter and filtrate were measured with a high-resolution germanium detector.

The distribution coefficient (K_d) was calculated as:

$$K_d = \frac{A_{SED}}{A_{FIL}}$$ (8.5)

where, A_{SED} is the net radionuclide activity measured on the sediment (Bq/g dry) after subtraction of the filter blank and A_{FIL} is the radionuclide activity measured in the seawater filtrate (Bq/g water).

Radionuclide adsorption kinetics

In coastal environments, geochemical, physical and sedimentological processes occur on time scales that are far shorter than in the deep-sea. Sorption reaction rates for many radioactive and non-radioactive contaminants occur over similarly short time scales (Li et al. 1984a; Nyffeler et al. 1984; Santschi et al. 1986; Jannasch et al. 1988). Rapidly fluctuating coastal processes are likely to influence the outcome of interactions between sediments and radionuclide-contaminated seawater. Based on time profiles of radionuclide uptake for sediments from Abrosimov and Stepovogo Bay, adsorption to sediments is rapid for ^{110m}Ag and ^{134}Cs with near-equilibrium achieved within one day after the start of the experiment (Fig. 8.11).

A more gradual increase in the distribution coefficients with time occurred for all other radionuclides. The results of the kinetic experiment imply that if the contact time between sediment-laden seawater is on the order of days or less in the bays, the dispersion away from the dumping sites of released radionuclides in the soluble fraction will be greater than predicted by equilibrium models of particle scavenging. Conversely, radionuclide exposures caused by local transport of sediment would be overestimated using an equilibrium distribution coefficient.

Influence of salinity

For the salinity experiment, water of different salinities (0‰, 5‰, 15‰, 25‰ and 34‰) was prepared by diluting Abrosimov Bay seawater (pH = 7.8; salinity = 34‰) with pure mineral water (pH = 7.6). Sediment concentrations in bottles were 55.0 mg dry/l for Stepovogo Bay sediment and 54.3 mg dry/l for Abrosimov Bay sediment. Notable changes in distribution coefficients are observed at low salinities (0-10‰), while only minor changes occur at higher salinities (Fig. 8.12). With the exception of ^{57}Co-cobalamine, distribution coefficients decrease as salinity increases. Changes in distribution coefficients with salinity demonstrates the often observed non-conservative behaviour of elements in estuaries where freshwater and seawater first mix (Whitfield and Turner 1979; Li et al. 1984b; Santschi et al. 1986 and others, also see Chapter 6). Comparison of salinity ranges measured in the two bays in Aug-Sept 1994 (31.5-34.1‰ in Abrosimov Bay and 17-35‰ for Stepovogo Bay) with the results of our experiments suggests only minor shifts in the equilibrium distribution coefficients will occur as a result of salinity fluctuations in the bays.

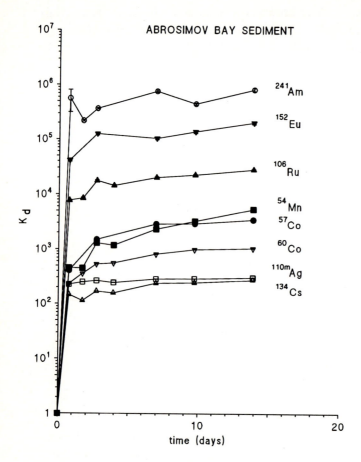

Fig. 8.11. Kinetic profiles of adsorption distribution coefficients for radionuclides (Carroll et al. 1996).

Sediment characteristics

An additional finding from the salinity experiment was that there were notable differences in distribution coefficients for ^{241}Am, ^{60}Co, and between sediments from the two bays. For Abrosimov Bay, the ^{109}Cd distribution coefficient was below the threshold of detection ($K_d < 10$). Adsorption distribution coefficients increased by a factor of five for ^{241}Am, six hundred for ^{60}Co and forty for ^{109}Cd (at 34 ‰), between Abrosimov Bay and Stepovogo Bay sediments. Higher distribution coefficients in Stepovogo Bay may be related to differences in sediment grain size. The mean particle diameter of Stepovogo Bay sediment (6.0 μm) is approximately one-half the mean particle diameter of Abrosimov Bay sediment (13.1 μm). Other differences in the characteristics of the sediments such as the base exchange capacity or clay mineral content are also likely to play a role. Such large differences in distribution coefficients between sediments from two different locations represents the main source of variation for distribution coefficients determined in these experiments.

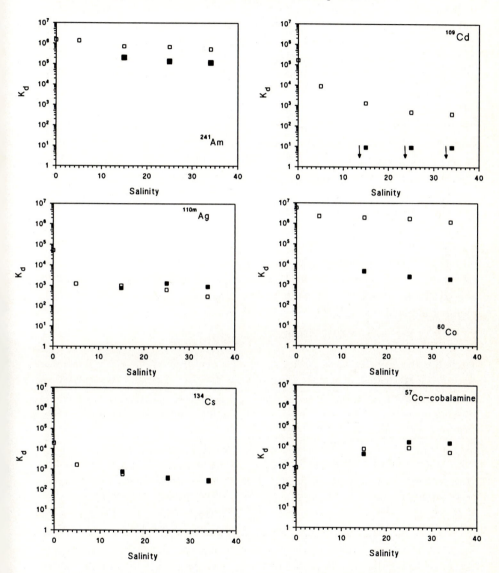

Fig. 8.12. Variation in distribution coefficients as a function of salinity (‰) after 14 days contact time (Stepovogo Bay sediments □; Abrosimov Bay sediments ■) (Carroll et al. 1996).

Influence of sediment concentration

Particle concentrations in the bays are likely to fluctuate within the range typically measured in coastal bays; i.e. 10's to 100's of mg/l with an increase with water depth. Such large variations are expected to play a significant role in the sorption behaviour of radionuclides. This hypothesis was tested using sediment from Abrosimov Bay only. Five concentrations of sediment were added to bottles containing 34‰ seawater: 0.54, 5.43, 10.86, 54.3, and 543 mg dry/l. A dependence of sorption behaviour on particle concentration has been observed previously over a seven-order of magnitude range in particle concentrations (Honeyman et al. 1988).

Fig. 8.13 shows that several radionuclides are sensitive to sediment concentration changes. The radionuclides ^{241}Am, ^{57}Co, ^{106}Ru, ^{85}Sr, and ^{133}Ba exhibit slopes of sediment concentration versus distribution coefficient significantly different from zero ($p < 0.001$). For the radionuclides ^{85}Sr, ^{133}Ba, and ^{57}Co-cobalamin, distribution coefficients vary with sediment concentration by as much as three orders of magnitude. In contrast, ^{241}Am and ^{106}Ru are less influenced by sediment concentration while ^{134}Cs, ^{60}Co, ^{152}Eu and ^{110m}Ag are not influenced by sediment concentration. The exact cause of the relationship between sediment concentration and sorption behaviour has not been determined; however, Honeyman et al. (1988) argue that differences among elements can be adequately explained based on surface coordination and colloid chemistry scavenging models. The range of K_ds determined for each element is given in Table 8.6.

Table 8.6. Range of distribution coefficients determined in laboratory studies. The comments represent the radionuclide used in the experiments and the types of experiments conducted.

	K_ds (g/ml)		
Nuclide	Minimum	Maximum	Comments
Cobalt	1.2×10^3	1.9×10^6	^{60}Co; salinity, sediment concentration
Strontium	15	130	^{85}Sr; salinity, sediment concentration
Europium	1.2×10^5	1.6×10^5	^{152}Eu; sediment concentration
Americium	1.1×10^5	2.6×10^6	^{241}Am; salinity, sediment concentration
Caesium	190	760	^{134}Cs; salinity, sediment concentration
Ruthenium	1.8×10^4	4.0×10^4	^{106}Ru; sediment concentration

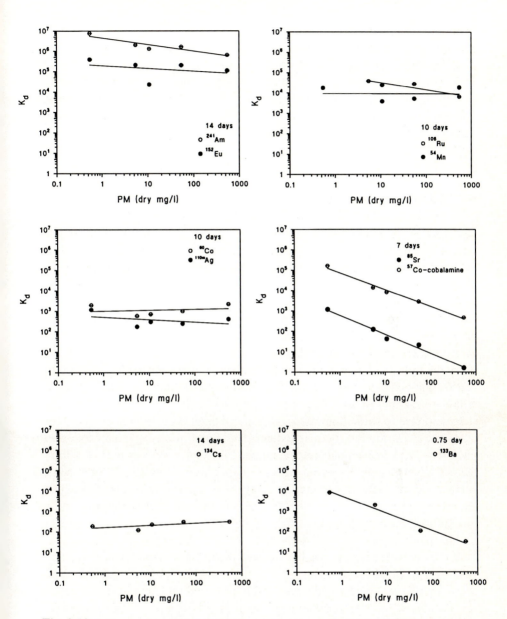

Fig. 8.13. Influence of suspended sediment concentration on distribution coefficients. Contact times vary between 18 hrs and 14 days depending upon the radionuclide (Carroll et al. 1996).

Comparison of laboratory K_ds with IAEA estimates

Unique characteristics of sediments found at dumping areas in the Kara Sea resulted in different K_d values than those reported in IAEA (1985). However, with the exception of ^{57}Co-cobalamine, ^{60}Co and ^{106}Ru, the distribution coefficients determined in this study overlap with the lower range of the values recommended for coastal areas by IAEA (1985) (Fig. 8.14). Exceptions were ^{60}Co with a distribution coefficient in Abrosimov Bay below the IAEA (1985) recommended range and ^{106}Ru which was above the IAEA (1985) range. Differences between the two data sets exemplify the value of conducting site-specific field investigations of K_ds when K_ds are to be used for waste management decisions.

8.6.5 Field experimentation

A drawback of conducting laboratory studies is changes begin to occur to PM and seawater as these environmental materials age. Experiments must be conducted as soon as possible after sample collection. This limitation poses difficulties when conducting research in remote areas such as in the Arctic. A second limitation of laboratory investigations is that typically only a few water and sediment samples are available from a particular region. Often, experiments must be conducted with sediments and water collected from different sites. It is not uncommon for laboratory experiments to be conducted with PM from the primary region of interest but with water from a completely different ocean. This severely limits the extrapolation of distribution coefficients from a limited number of sites to a large region.

In response to these limitations, an experimental protocol was designed for conducting distribution coefficient determinations at sea using freshly collected materials. This investigation was conducted in August 1995 as part of a Joint Russian/Norwegian/American Military Expedition. Ship-based experiments were conducted under a radiation safety plan approved by the Norwegian Radiation Protection Authority. The approved safety plan included strict precautions on every aspect of the handling of radioactivity and processing of radioactive samples onboard the vessel.

Batch experiments were conducted using the tracers ^{241}Am, ^{57}Co (inorganic), and ^{134}Cs, radiotracers with varying degrees of particle affinity. The experiments were performed using freshly collected sediment and water from 25 stations (Fig. 8.1) representing the primary regions of the Kara Sea. Sediments (upper 2 mm) were retrieved from box cores and mixed with filtered bottom water from the same station. Sediment concentrations in the experimental bottles ranged from 1-70 mg dry/l. The experimental design was similar to that used in the laboratory experiments except that the entire sample (250 ml) was filtered five days after the addition of the radionuclides.

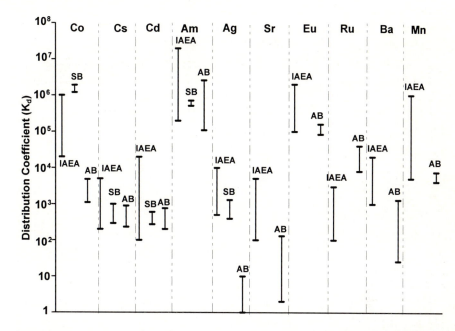

Fig. 8.14. Comparison of K_d ranges published by the International Atomic Energy Agency (IAEA,1985) with ranges determined for sediments from nuclear waste dumping sites in Abrosimov Bay and Stepovogo Bay.

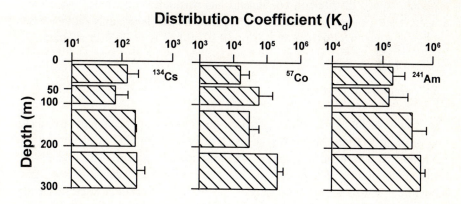

Fig. 8.15. Distribution coefficients as a function of water depth for ^{134}Cs, ^{57}Co and ^{241}Am using surficial bottom sediments.

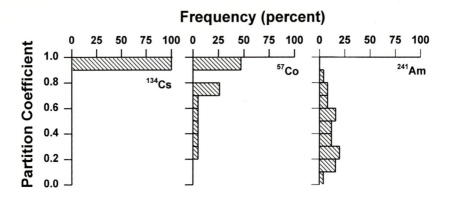

Fig. 8.16. Partition coefficients determined in field experiments for ¹³⁴Cs, ⁵⁷Co and ²⁴¹Am using surficial bottom sediments.

The data show a trend of increasing distribution coefficients with water depth for the radionuclides ²⁴¹Am and ⁵⁷Co (Fig. 8.15). This result is due to a combination of the processes presented in Table 8.3 which differ between the deep basin (Novaya Zemlya Trough) and shallow shelves in the Kara Sea.

Partition coefficients may be calculated directly from the experimental data (Fig. 8.16). These data confirm the finding of the theoretical and laboratory experiments that distribution coefficients and partition coefficients both exhibit a wide range of variation for the Kara Sea. Caesium is the only radionuclide exhibiting a well-constrained K_d (0.14±0.02 m³/kg) with 99% in solution.

8.6.6 Summary of findings

The results of the three-phase approach to determining K_d values and their ranges of uncertainty may now be used to re-evaluate PC_w values used by modellers to determine the impact of future waste releases. The probability distribution of PC_w determined by computer simulation demonstrated that most of the waste-related radionuclides are expected to exhibit large ranges of variation for Kara Sea PC_w values. The only exceptions are radionuclides with low K_ds, i.e. caesium and strontium. With the exception of americium, ranges for all of the other radionuclides span nearly the entire range of partition coefficient values. Laboratory experiments identified PM characteristics as the major source of variation for K_ds. A comparison of K_d values determined at sea from 25 locations confirmed that K_d values vary considerably among locations. Partition coefficients calculated directly from the field data also demonstrate that PC_w will vary considerably among locations in response to low availability of PM for sorption. The analysis demonstrates the necessity of conducting sensitivity tests on radionuclide partition coefficients as part of modelling exercises for the Kara Sea. Although it is much easier for modellers to assume average K_d and PM values as representative of the Kara Sea, such an approach oversimplifies the environment and may lead to erroneous conclusions concerning contaminant mobility.

8.6.7 Models for predicting radionuclide dispersion

Box and hydrodynamic computer models are used to evaluate radionuclide transport in major regions of the Kara Sea over varying time scales. Model assumptions and input requirements for the two model types are different. The primary difference between the two approaches is in the level of spatial and temporal detail given for the regions represented in the models. Box models contain little structural detail while hydrodynamic models contain a detailed representation of oceanographic characteristics. As a result, box models are suited for conducting assessments of the impact of radionuclide releases over long time scales (greater than 100 years) and large regions while hydrodynamic models are more useful for high resolution, short term predictions (years to decades).

Compartmental models
The class of models known as box, or compartmental models represents individual areas as compartments which exchange water and contaminants with time. Once the total number of compartments, their arrangement and interactions have been established, the primary input requirement of these models is the water flux (m^3/s) in and out of each compartment and, of course, contaminant discharges.

The transfer of water and radionuclides between compartments is governed by the basic physical law known as *conservation of mass*. The general form of the differential equation describing the concentration of radionuclide in compartment i is given as (Sazykina and Kryshev 1994):

$$V_i \frac{dC}{dt} = \sum F_{ji} C_j - \sum F_{ij} C_i - k_i C_i V_i + Q_i \qquad (8.5)$$

where
C_i = radionuclide concentration in water compartment i ;
V_i = volume of compartment i ;
F_{ij} = fluxes from compartment i to compartments j $(j = 1,n; F_{ii} = 0)$;
F_{ji} = fluxes from compartment j to compartment i ;
k_i = loss of activity in the compartment, e.g. radioactive decay, sorption
Q_i = sources of radioactivity in compartment , e.g. releases from dumping
grounds

The parameter k_i represents the fraction of the total radioactivity in the box lost as a result of removal mechanisms including radioactive decay and the uptake to sediments as a result of sediment-water partitioning. Uniform mixing is assumed for water within each box and the distribution of radioactivity is also assumed to be uniform. Many models also include a sediment deposition term to trace the loss of contaminated sediment from the water column to the seabed.

The disadvantage of compartmental models is that they do not contain any representation of bottom topography or water column structure, both of which greatly influence transport and mixing of individual water masses within a specific region of the ocean. Where detailed information on water mass transport and mixing processes is poorly known anyway, such models can provide an understanding of the general movement of contaminants within major regions of the Arctic.

One of the compartment models presently in use for modelling the dispersal of radioactive pollutants is the 16 box ARCTIC-2 model (Hamilton et al. 1994) (Fig. 8.17). This model has four compartments representing the Kara Sea: Kara Sea West, Surface and Bottom and Kara Sea East, Surface and Bottom with flows to and from the Barents Sea, and flows to the Laptev Sea, the Arctic Ocean, and the Svyatata Anna Trough. The model includes a loss term associated with sedimentation whereby contaminated sediments suspended in the water column are incorporated into the seabed. The model produces radionuclide concentration data in the different compartments as output.

A contaminant release scenario of gradual release of ^{137}Cs over 20 years (continuous release) following dumping of naval reactors containing spent nuclear fuel is presented for the ARCTIC-2 model in Fig. 8.18. The model predicts maximum concentrations of 30 Bq/m^3 from the Kara Sea West compartment.

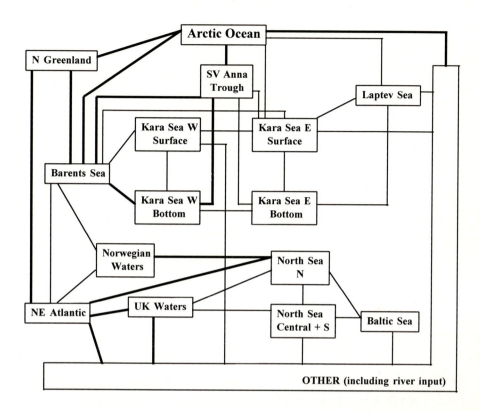

Fig. 8.17. Structure of the ARCTIC-2 compartmental model (from Hamilton et al. 1994).

   This corresponds to less than 1% of the natural radioactivity of seawater and suggests that if such a release scenario is accurate, little if any environmental impact would be observed on a global-scale as a result of releases from within the Kara Sea.

Hydrodynamic models

Three-dimensional hydrodynamic models are based on highly sophisticated numerical schemes used to reproduce in as much detail as possible, the spatial and temporal variability of processes operating in oceans. Bottom topographic changes are represented over scales of 10's of km and the water column is typically modeled as several horizontal layers of 10's to hundreds of meters in thickness. The dynamics of ocean circulation operating within the topographic boundaries are governed by the basic laws of physics. These laws are (Pond and Pickard 1983):

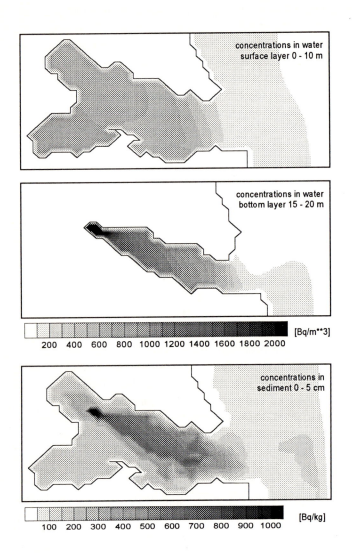

Fig. 8.19. Stationary state concentrations of [239]Pu in Abrosimov Bay for a continuous release scenario of 1TBq/year from within Abrosimov Bay (Harms 1995).

Modelling simulations are underway to assess the potential radionuclide flux to the Kara Sea should a catastrophic release of radionuclides occur and to reconstruct past radionuclide fluxes.

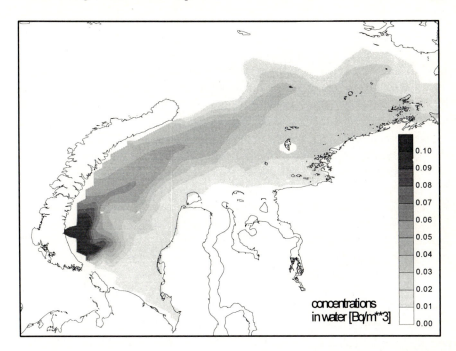

Fig. 8.20. Annual average concentrations (year 6) of ^{239}Pu in the Kara Sea for a continuous release scenario of 1TBq/year from Abrosimov Bay (Harms 1995).

8.7 Impact assessment

As previously demonstrated, the main pathways and concentrations of radionuclides in the Kara Sea have been discerned by combining information on waste inventories with modelling exercises. The final step of the risk assessment is to assess the impact of the dumped wastes on global, regional and local populations. This analysis is conducted by identifying the primary ways whereby humans are exposed to radioactivity and quantifying the amount of exposure from each source. Of primary concern is the effect of ingestion of contaminated seafood harvested from the Kara Sea. Biological communities living in the pathway of contaminated water and sediments are first identified. Estimates are then made of the amount and type of organisms likely to be ingested by humans from the contaminated regions (e.g. fish and shellfish). The expected concentration levels of contaminants in the organisms are quantified (see section 2.2.3.1) and the impact on humans of ingesting contaminated organisms is assessed.

Compared to the Barents Sea, the Kara Sea is far poorer in biomass and species diversity. Ice covers the Kara Sea for nearly three quarters of the year and primary production is limited. Benthic biomass ranges from approximately 3 to 10 g/m^2 and the fish population is very small (Fowler et al. 1994). As a result, there is no commercial catch of fish in the central Kara Sea (Miquel 1996). With low productivity and few small sized fish of commercial value, the probability of contaminant transfer to higher trophic levels through the food web in the central Kara Sea is small (Fig. 8.21).

Commercial fishing does occur in the shallower waters of the bays of Novaya Zemlya and estuaries of the Ob and Yenisey Rivers where biomass increases to 100-300 g/m^2 (Fowler et al. 1994). However few reliable statistics exist on fish catches for the Kara Sea except for the estuaries of the Ob and Yenisey. In 1990 the total catch of fish in the Ob estuary was 1526 tonnes of sea and brackish water fish and 836 tonnes of freshwater fish. For the Yenisey the catch was 228 tonnes (Sazykina and Kryshev 1994; Miquel 1996). Whitefish were the most abundant species in both estuaries (Miquel 1996). Hence dose commitment estimates for man from food are performed considering sources of contaminants in the coastal waters and bays and nuclear and non-nuclear pollution from the rivers.

While a significant body of literature exists on contaminant uptake to marine organisms in temperate and tropical ecosystems, few studies have previously been conducted on the metabolic effects of cold temperatures on the uptake and mobilization of contaminants by biological organisms (Boisson et al. 1996; Hutchins et al. 1995).

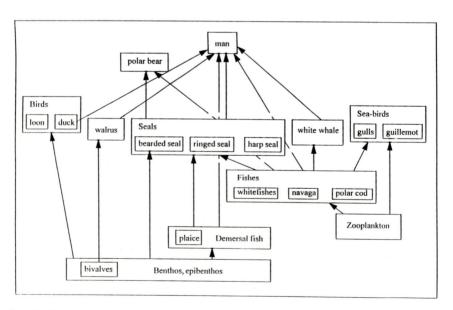

Fig. 8.21. Trophic food web for the coastal Kara Sea (Miquel 1996).

Although more information is needed on this subject, it appears that presently available data are acceptable for purposes of conducting the global and regional scale risk assessments for the Kara Sea.

In experiments conducted on organisms from northern habitats, temperature affects were observed for metabolic elements but most radionuclides associated with waste dumping show no relationship with temperature. For the brittle star *Ophiothrix fragilis,* Hutchins et al. (1995) observed little temperature effect for primary waste related radionuclides while Boisson et al., 1996 observed a temperature affect for ^{60}Co for the marine alga *Fucus vesiculosus* (L.).

8.7.1 Global impacts

An assessment has been conducted on a global scale using the Arctic-2 model for a scenario of a single release of ^{137}Cs in each of the Kara and Barents Seas (Hamilton et al. 1994). The results show that up to 3 and 7 manSv are delivered to the world population through fish ingestion. For release in the Barents Sea, over 50% of the dose commitment is delivered via Barents Sea fisheries, the remaining part coming from the northeast Atlantic (13%) and the rest of the world ocean (30%). A smaller dose commitment (25%) is associated with a release from the Kara Sea as a consequence of the absence of large-scale fishing. For a continuous release scenario in which ^{137}Cs is released gradually over 20 years, the dose commitment is negligible.

These dose commitment estimates are low when compared to other sources of contamination to human populations (Table 8.7) and thus there is little cause for concern over the potential spread of contamination resulting from the wastes currently residing in the Kara Sea.

8.7.2 Regional impacts

On a regional scale, populations most subject to risks are inhabitants of the coastal regions of Taymyr and Yamal (Osvath et al. 1995). An assessment was conducted using hydrodynamic models of circulation and radionuclide dispersion (Povinec et al. 1996; IAEA-MEL 1994; Harms 1992; Harms and Backhaus 1992). A simulation was conducted assuming an instantaneous release of 1 PBq of ^{137}Cs from all nuclear reactors dumped in the Kara Sea. Such a release produced concentrations of less than 15 Bq/m^3 in the Taymyr and Yamal coastal regions corresponding to a maximum individual dose below 5 μSv/yr from fish ingestion (Povinec et al. 1996).

Table 8.7. Collective dose commitment data (Osvath et al. 1995; IAEA-MEL 1994). On a global scale the radiological impact is negligible, with a collective effective dose commitment from seafood ingestion of about 10 manSv.

Source	Dose (manSv)
Collective exposure of the global population	
Nuclear weapon tests fallout	3×10^7
Natural radionuclides in seawater	1.8×10^7
One year nuclear fuel cycle	8×10^4
Arctic dumping of radioactive wastes	10
Collective exposure of the population of the European Community from	
radionuclides in the northern European waters (MARINA)	
Natural radionuclides	1.7×10^6
Civil nuclear site discharges	5300
Nuclear weapon tests fallout	1600[#]
Chernobyl fallout	1000
Northeast Atlantic dumping of radioactive wastes	50

[#] ^{14}C contribution not included

For a gradual release of 1 TBq/yr of ^{137}Cs from each of Abrosimov, Stepovogo and Tsivolki Bays and the Novaya Zemlya Trough, concentrations were estimated to be below 0.5 Bq/m^3 at the Taymyr coast, corresponding to a maximum individual dose below 0.2 μSv/yr (Povinec et al. 1996). As was seen on a global scale the predicted levels of contamination pose little threat to regional populations subsisting on fish from the Kara Sea.

8.7.3 Local impact

On a local scale, experiments were conducted using hydrodynamic models developed for Stepovogo and Abrosimov Bays (Povinec et al. 1996). Simulations were conducted assuming ^{137}Cs inventories in the bays of 0.13 PBq and 0.36 PBq. For a release of 1 TBq/yr from the dumped waste, ^{137}Cs concentrations in water inside the bays were determined to be on the order of 3 kBq/m^3. Given these conditions the maximum individual dose to a hypothetical individual consuming fish from one of the bays is predicted to be below 1 mSv/yr (Povinec et al. 1996). Because 1 mSv/yr is the acceptable dose limit for the public, additional site-specific information will be collected to help further evaluate the radiological effects on a local scale. The results of the dose estimations for local, regional and global scales are summarized in Table 8.8.

Table 8.8. Maximum individual dose rates from dumped radioactive wastes in the Arctic Seas at the time of dumping (IAEA-MEL 1994).

Dose type	Dose
Global	3 μSv/yr
Regional (per TBq/yr release)	1-10 μSv/yr
Local (per TBq/yr release)	10-100 μSv/yr

8.7.4 Conclusions

The presence of radioactive waste in the Kara Sea has elicited concern from government representatives from many nations. This problem has been addressed through the development and implementation of a step-wise risk assessment under the auspices of the International Atomic Energy Agency with participation from experts from many nations. The components of the program were presented with major emphasis on the use of distribution and partition coefficients in radionuclide dispersion models. Techniques for characterizing the magnitudes and uncertainties for K_ds and PC_ws and the limitations of the techniques were presented. The approaches demonstrate the necessity of conducting sensitivity tests on radionuclide partition coefficients as part of modelling exercises for the Kara Sea.

The presence of high concentrations of a variety of radionuclides in this shallow coastal sea is obviously undesirable. However, the waste containers do not appear to present an imminent hazard. Present and past environmental levels of radionuclides are not appreciably elevated above expected levels and modelling investigations of waste release scenarios suggest there is little cause for concern over the global spread of radioactivity in the marine environment.

The primary exposure route for humans is through the consumption of seafood from the Kara Sea. On a global scale the radiological impact on humans is negligible. Estimates of the dose commitment on a regional scale also appear to be minimal while there remains some question of local scale risk.

In view of the difficulties of devising cost-effective and safe strategies for retrieving the wastes from the seabed, it is likely that a no-action remediation strategy will be followed in the Kara Sea. Long-term monitoring with permanent monitoring stations near the dumpsites to detect future releases is a feasible strategy for the near term. However, further study of the dumpsites with respect to potential impacts on a local-scale are likely to be recommended so that a better understanding of this risk may be pursued.

9 Global oceanic and atmospheric stability of oxygen

9.1 Oxygen budgets

9.1.1 Global budgets

Prehistoric atmospheric oxygen production began through water vapour dissociation, caused by powerful UV radiation from space, followed by a secondary reaction involving the formation of ozone (O_3) from O_2. The level of oxygen, called Urey level, was kept stable by the feedback process where UV radiation is reduced through absorption by the ozone produced. The equilibrium level thus obtained is estimated at 1 ‰ of the actual level, i.e. about 0.02 vol%. At 3.2 billion years ago an initial form of primitive photosynthesis started, keeping the oxygen level at about 0.2 vol% (Pasteur level), followed at 1.8 billion years ago by photosynthesis using chlorophyll pigments (Duursma and Boisson 1994). Photosynthesis started using chlorophyll pigments. This occurred first in water systems below a column of about 10 m water, out of reach of the still high level of hazardous UV radiation, which was absorbed by this water layer. Later, when a sufficient level of atmospheric O_2 was built up, and consequently of an increased ozone level, this ozone absorbed the major part of hazardous UV radiation. As a result, photosynthesis manifested itself in surface waters and on land.

A net amount of 5.63×10^{20} mol (1.8×10^{22} g) is produced, of which 0.375×10^{20} mol is present as free oxygen in the atmosphere, 3.1×10^{17} mol as dissolved oxygen in the oceans, while the rest is stored in a great number of oxidized terrestrial and oceanic compounds (Fig. 9.1).

Since atmospheric oxygen has been produced due to burial of organic matter, a similar quantity of organic carbon is needed to balance the total oxygen produced (Fig. 9.1). Stored palaeo-oxygen in sulphur oxides can be found in oceanic SO_4^- (concentration 2.7×10^3 g SO_4^-/m^3 containing 56.25 mol O_2/m^3), which contributes 0.78×10^{20} mol O_2. This represents an O_2 reserve twice above that of the total atmospheric O_2. Forms like oceanic NO_3^- (2-5 g N/m^3 containing 0.22-0.42 mol O_2/m^3) are less significant in this respect. The remaining stored oxygen is to be found in sedimentary sulphur and metal oxides, in particular iron oxides (Luther 1990; Luther et al. 1992).

The amount of organic carbon, buried in oceanic and terrestrial sediments, is estimated by Degens (1982) to be 3.1×10^{20} mol Org.-C, of which 3×10^{20} mol is present in shales. By comparison, the world coal and oil reserves are only 5×10^{17} mol C and 1.7×10^{16} mol Org.-C, respectively.

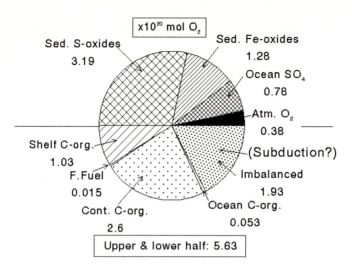

Fig. 9.1. Pie chart of the budget of historically produced oxygen by photosynthesis (upper half) over 3.2×10^9 years. The lower part concerns fossil organic matter, thus produced, expressed in oxygen equivalents. Data are given in 10^{20} mol O_2.

Supposing that each buried organic carbon atom requires 1-2 molecules of O_2, the oxygen equivalent is $3.1\text{-}6.2 \times 10^{20}$ mol O_2, which is a factor of 8.2-16.4 times higher than the present atmospheric and oceanic oxygen reserve ($= 3.78 \times 10^{19}$ mol O_2; Table 9.3). Budyko et al. (1987) confirmed this range of buried organic carbon as given by Degens (1982; 1989), to be equivalent to 3.70×10^{20} mol O_2 for continents, shelves, and ocean floors (Fig. 9.1), which is an order of magnitude higher than the atmospheric oxygen mass, but lower than the total produced oxygen (5.63×10^{20} mol).

The origin of the imbalance (1.93×10^{20} mol O_2) may be organic matter originating from the ocean and shelf bottom which has disappeared under the continents by subduction. The total sediment subduction is estimated at $2\text{-}6 \times 10^{15}$ g/yr, which leads Fyfe (1992) to conclude that, given the mass of the continental crust of 1.6×10^{25} g, the process of full recycling occurs once every 3×10^9 years. This provides some support for subduction as the cause of the imbalanced organic matter. However, since Fe and S oxides are equally involved in subduction, further speculation is not warranted without the availability of a better set of data.

A recent estimate however gives a much higher value for sedimentary rock organic matter (Keeling et al. 1993). Their value of 1×10^{21} mol of organic-C is far higher than that of Budyko et al. (1987) and twice the total atmospheric and 'stored' oxygen of 5.63×10^{20} mol, given in Fig. 9.1. The wide variation in estimates by different scientists points to the need for an updated research effort to provide reliable data, both for sedimentary organic matter, fossil fuel and charcoal, as for estimates on sedimentary sulphur-oxides and iron-oxides.

Models show that atmospheric oxygen during the last 600 million years ranged between 7 and 30 O_2 vol% (one model even up to 35 O_2 vol%) (Fig. 9.2). The level of oxygen became finally 20.946 O_2 vol% in more recent times.

The major feedback systems, which could stabilize oxygen at 20.946 vol% over time scales of 1000's of years are not known, but should be related to the balance between photosynthesis (production of oxygen) and respiration (use of oxygen) on the one hand and to the oxidation of reduced products like Fe(II) compounds on the other (Kump and Holland 1992). The process that causes additional oxygen is the burial of organic matter, while weathering of buried organic matter will again consume O_2. On the basis of rates of burial and weathering of organic carbon, Budyko et al. (1987), Berner and Canfield (1989) and Kump (1992; 1993) came to conflicting modelling results on oxygen fluctuations for the period of 600 million years ago until the present day. Their differences mainly concern the periods of maximum and minimum atmospheric O_2. The basis for their models is given by Garrels and Perry (1974) in equation (9.1).

$$15CH_2O+8CaSO_4+2Fe_2O_3+7MgSiO_3 \rightleftharpoons 8CaCO_3+4FeS_2+7MgCO_3+7SiO_2+15H_2O$$
$$(9.1)$$

This formula indicates a linkage of oxidized reservoirs ($CaSO_4$, Fe_2O_3, $CaCO_3$) and reduced reservoirs (FeS_2, CH_2O), where CH_2O is organic matter. The formula is the basis for the BLAG model, worked out by Berner et al. (1983).

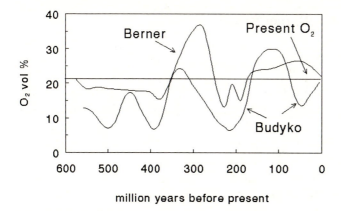

million years before present

Fig. 9.2. Response of total atmospheric oxygen from 600 million years until present, as modelled by Budyko et al. (1987) and Berner and Canfield (1989).

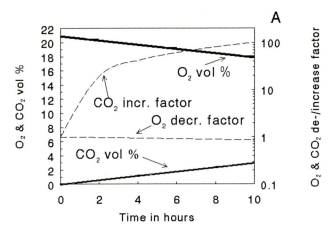

Fig. 9.5A. Oxygen consumption and CO_2 production of 10 persons in a 50 m^3 space (Duursma and Boisson 1994).

On the other hand hypercapnia may occur due to elevated CO_2 levels (Fig. 9.5C). After 10 hours the CO_2 level has augmented to 101 times of its original value, resulting in a partial pressure of pCO_2 of 4 Torr (mm Hg). Although the human organism can tolerate a pCO_2 level in blood of 80 Torr for short periods, lower pCO_2's in the lungs of, e.g., 45 Torr will result in severe hyperventilation, migraine and stomach problems (Soulez-Larivière and Le Péchon 1991; ESA 1992) (Fig. 9.5C).

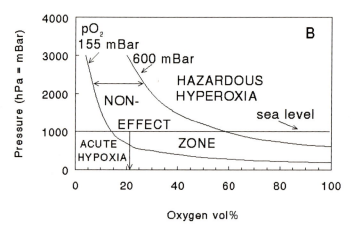

Fig. 9.5B. Oxygen effect zones on man after Le Péchon (ESA 1992)

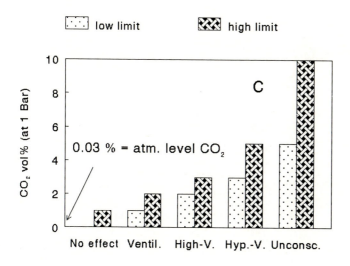

Fig. 9.5C. CO_2 effect zones on man after Le Péchon (ESA 1992). Ventil. = ventilation; High-V. = high ventilation; Hyp.-V = hyper ventilation; Unconsc. = unconsciousness.

The danger in Paris is the quantity of CO_2 produced and associated exhaust gases. The CO_2 produced is equivalent to an amount of CO_2 present in an equivalent atmospheric layer (all of 1000 mBar) of 16,850 m^3/m^2 (Fig. 9.4). Hence at periods of high atmospheric stability, CO_2 levels may rise to high levels, causing some of the nuisance effects mentioned in Fig. 9.5C. CO_2 effects are however preceded by those caused by exhaust gasses, like NO_x SO_2, O_3, CO and smoke, which play an overriding role and reach intolerable values in a number of cases (Table 9.1).

Table 9.1. Air quality of Paris in 1994 as determined for NO_2, SO_2 and O_3 (in $\mu g/m^3$), and threshold values. Min Environm (1995).

	NO_2	SO_2	O_3
Average 1994 surpassing threshold:	74	16	Max.: 274
	for 2% of time > 156-204		102 days > 180 0 days > 360
Thresholds (3 degrees of alert)	1: 200 2: 300 3: 400	1: 200 2: 350 3: 600	1: 130 2: 180 3: 360

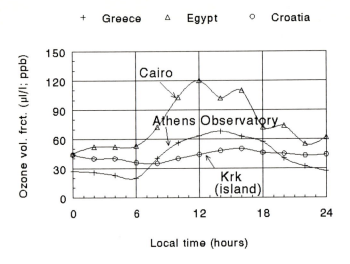

Fig. 9.6. Daily rhythm of ozone concentrations. Krk island (background), Cairo and Athens.

Photosmog results from reactions between nitrogen oxide(s), oxygen, hydrocarbons and other oxidizable molecules under influence of sunlight (Klasinc and Cvitas 1996). It is related to natural phenomena of ozone formation in the lower troposphere during daytime (Fig. 9.6) with a maximum in the afternoon and a recovery during the night. The threat in large towns is caused by the fact that the excess daytime production is not compensated for during the night, resulting in an elevated level of ozone during the day.

9.2 Relation to CO_2 budgets

9.2.1 Atmosphere and ocean budgets

On small time scales, oxygen is negatively correlated on a 1:1 basis to atmospheric CO_2, which has a level of 0.03 vol% or 300 ppmv (1965; 1992 level: 360 ppmv). Thus, even close to vegetation, atmospheric oxygen cannot increase more than this same amount (0.03 vol%) and can therefore not exceed 20.976 vol %. The chance of an increase of forest fires, due to oxygen increase, is negligible.

The atmosphere-ocean exchange budget of CO_2 is dominated by the fossil-fuel combustion and biomass fires. From the CO_2 liberated during this process, 39% remains in the atmosphere, 30% is taken up by the oceans, the rest possibly being recovered by biomass growth.

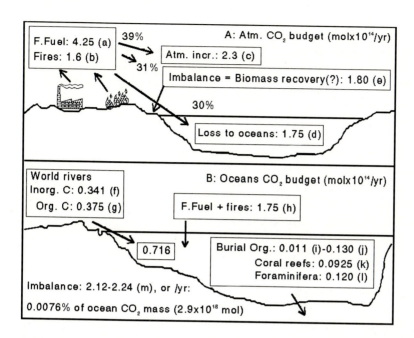

Fig. 9.7. CO_2 budgets in atmosphere (A) and oceans (B). For references, numbered (a) to (m) see Duursma and Boisson (1994).

Compared to the atmosphere-ocean flux of CO_2, the biogenic ocean fluxes of carbon ('oceanic biological pumps') are at least a factor 10 times lower (burial of organic matter and foraminifera, and coral reef growth) (Fig. 9.7B).

The fossil-fuel derived CO_2, which is released at a rate of about 4.25×10^{14} mol CO_2/yr will not rapidly equilibrate with the ΣCO_2 in the oceans (Quay et al. 1992). The residence time of water masses, which transport CO_2 into the deep sea in polar regions, where it is effectively sequestered, is of the order of a thousand years (650 year in the Atlantic to 2000 year in the Pacific according to Groen (1974)). Hence a disequilibrium between atmospheric and oceanic CO_2 will exist for several millenaries after all fossil-fuel reserves have been exhausted.

The dissolution process of CO_2 into the ocean basin will affect the CO_2 equilibrium in sea water and the pH by shifting the equilibria given in the reactions (9.3) (Buch et al. 1932; Wollast and Vanderborght 1994).

$$CO_2 + H_2O \rightleftharpoons H^+ + HCO_3^- \quad (K_1)$$
$$HCO_3^- \rightleftharpoons 2H^+ + CO_3^{2-} \quad (K_2)$$

(9.3)

For CO_2 the exchange coefficient and net CO_2 fluxes from and to the oceans were determined world-wide by Etcheto and Merlivat (1988) and Merlivat et al. (1991) using the SEASAT satellite scatterometer data. CO_2 is absorbed by the oceans in the cold polar waters and released to the atmosphere in the tropics. The net CO_2 flux into the oceans is estimated at 0.975×10^{14} mol CO_2/yr (Etcheto and Merlivat 1988), while other literature data range from 0.74×10^{14} mol CO_2/yr (cited by Merlivat et al. 1991) to 1.75×10^{14} mol CO_2/yr (Quay et al. 1992) (Fig. 9.7A). These fluxes are considered to represent the fossil-fuel and biomass fires derived CO_2 from atmosphere to oceans. Since 5.85×10^{14} mol C is released annually as CO_2, 30% is absorbed by the oceans, taking only the data of Quay et al. (1992), 39% is accumulated in the atmosphere and 31% may be recovered by regrowth of biomass. Since savannah and rainforest fires are the major part of biomass fires (Artaxo et al. 1993), this explains part of the annual imbalance due to recovery of vegetation growth of these savannahs.

A still unsolved problem concerns the present-day CO_2 fluxes between atmosphere and oceans. In Fig. 9.7B a budget is shown, in which the input of land runoff is fully taken into account. There exists an imbalance for CO_2 exchange which is slightly higher (2.12-2.24×10^{14} mol CO_2/yr) than detected by satellite scatterometer data (1.75×10^{14} mol CO_2/yr). The question is however whether a full exchange equilibrium should exist, since the time scale for reaching an equilibrium between gasses exchanged between atmosphere and oceans is on the order of several hundred years. A disequilibrium is furthermore possible because the imbalance amounts to only 0.0076% of the total CO_2 mass of the oceans (Fig. 9.7B).

Another interesting point is the burial of 0.22-0.34×10^{14} mol CO_2/yr through various means such as organic matter (0.011-0.13×10^{14} mol/yr), coral reef formation (1990 value, 0.0925×10^{14} mol/yr) and net sedimentation of foraminifera of 0.12×10^{14} mol/yr, taking into account that foraminifera redissolve in the deep ocean below the carbonate compensation depth.

The net influx of fossil-fuel and biomass-fire derived CO_2 from atmosphere to oceans of 1.75×10^{14} mol/yr is a factor 2½ times higher than the river-derived CO_2. Since the atmosphere to oceans CO_2 influx data derive from two different methods ($^{13}C/^{12}C$ evidence and satellite determination of the CO_2 air-sea exchange coefficient over complete oceans), the budget calculation in Fig. 9.7B is still somewhat rough, at least for the ocean margins. This may be related to CO_2 degassing from coastal waters or burial of carbonates in coastal sediments, which escape $^{13}C/^{12}C$ and satellite measurements. This shows that the air-sea processes occurring in the continental margins cannot be neglected, and may have world-wide impacts on global change processes. Also Sundquist (1993) recognizes this uncertainty in CO$_2$ budgets due to unidentified terrestrial CO$_2$ sinks.

9.2.2 The Antarctic paradox

The question is often posed whether atmospheric CO$_2$ uptake by the oceans can be enhanced in order to limit the steady atmospheric CO$_2$ rise and as such reduce the greenhouse effect. The primary productivity in the oceans, although mostly limited by available nutrients, was found to be limited by iron deficiency (Martin and Fitzwater 1988; Martin et al. 1990), at least in Antarctic waters where most of the CO$_2$ flux to the oceans occurs (Merlivat et al. 1991). Since nutrients in these waters were not exhausted during phytoplankton growth, the controversy which was created over this subject was called the Antarctic paradox.

Trace-metal deficiencies for photosynthesis are for many decades well known in agriculture for metals such as molybdenum, zinc and others (Kabata-Pendias and Pendias (1984); Foth and Ellis 1988). In response, trace-metals are added to fertilizers in order to overcome such a deficiency. This idea was extended to the oceans in order to potentially increase the transfer of atmospheric CO$_2$ to the deep ocean by the 'biological pump'. However, the idea failed to resolve the problem quantitatively (Broecker 1990; Watson et al. 1994). Also the extrapolation of experimental results of iron fertilization of phytoplankton cultures to nature requires careful attention. Dugdale and Wilkerson (1990) concluded from a re-evaluation of their ^{15}N uptake data in the Antarctic, that Fe addition does not affect algal growth.

Broecker (1990) argues that fertilizing marine waters with iron faces the problem of short residence time due to nonbiologic pathways such as hydrolysation and adsorption to particulate matter, which reduces considerably the availability of Fe ions to phytoplankton. To this can be added that addition of iron to sea water will potentially enhance the precipitation of other trace metals, being adsorbed to Fe(OH)$_3$ flocculates.

Watson et al. (1994) demonstrated with an *in situ* experiment over 64 km^2 in the equatorial Pacific Ocean, that the addition of iron revealed only a small effect on the CO$_2$ utilisation and drawdown of carbon ($< 10\%$), than what would have occurred had the enrichment resulted in the complete utilization of all available nitrate and phosphate. This experiment shows a clear difference with results obtained under on board laboratory conditions.

As mentioned in the former section 9.2.1, changes in atmospheric-oceanic CO$_2$ ratios will only be enhanced when the burial rate of organic matter changes significantly. It was already concluded (Fig. 9.7B) that a 10% change in burial rate cannot have a great effect ($0.02\text{-}0.034 \times 10^4$ mol/yr), since this is only 0.9-1.5 % of the existing imbalance (2.18×10^{14} mol/yr). Nevertheless, iron fertilization may have played a role in modulating atmospheric CO_2 levels between glacial and interglacial times (Watson et al. 1994). The iron hypothesis might therefore change in the process, but its central message is likely to survive (Cullen 1995).

9.3 Time-dependent processes

An increased world biomass (1.8×10^{17} mol CO_2; Duursma and Boisson 1994), resulting from a hypothetical increase in photosynthesis over respiration 1% would result in 1.8×10^{15} mol CO_2 being removed from the atmosphere and ocean CO_2 pool. Compared to the CO_2 annually released from fossil-fuel burning of 4.25×10^{14} mol CO_2 (Quay et al. 1992) this shows that the biosphere is a potential accumulator reservoir of fossil-fuel CO_2. However, an increase of photosynthesis in relation to respiration can only occur when more biomass and humus are formed permanently. Conversely, the present deforestation at a world scale contributes $0.8\text{-}2.5 \times 10^{14}$ mol CO_2/yr (Houghton et al. 1991) to the atmosphere which is rather the réverse of humus formation. A more precise figure for deforestation and biomass burning, deducting reforestation and biomass regrowth, is 0.52×10^{14} mol CO_2/yr (Bouwman et al. 1992).

An increased CO_2 uptake due to humus formation is less probable in the oceans. When we compare the atmosphere-ocean CO_2 flux of 1.75×10^{14} mol CO_2/yr (Fig. 9.7B) with that of the burial rate of organic carbon of $0.011\text{-}0.13 \times 10^{14}$ mol 'CO_2'/yr, we see that a 1 % change in photosynthesis over respiration should only cause a change in burial rate of organic matter of $0.00011\text{-}0.0013 \times 10^{14}$ mol CO_2/yr, which is 0.006-0.07% of the CO_2 air-ocean flux. It is virtually inconceivable that the ocean primary/respiratory production system ('oceanic biological pump') would have any measurable effect on fossil-fuel carbon dioxide fluxes from atmosphere to the oceans.

Feedback processes, regulating atmospheric oxygen in relation to photosynthesis-respiration rates, are difficult to prove for short periods of 100's or 1000's of years, but may be linked to shifts in C_3/C_4 photosynthetic activities both on land and in the sea. C_3 and C_4 photosynthetic pathways react differently on changing environmental conditions such as light, temperature, moisture (on land), CO_2, O_2 and nutrient levels. C_4 photosynthesis is more efficient than that of C_3 with respect to nutrients. At higher temperatures, C_4 photosynthesis is favoured on land, particularly in arid regions.

In the following section, the environmental feedbacks on oxygen regulation are discussed through time scales concerning the earth, ocean, ocean margin, glacial and interglacial periods. The potential contribution by the oceans in regulating variations in atmospheric oxygen is also evaluated.

9.3.1 Environmental effects on photosynthesis and respiration

In order to understand processes that regulate oxygen production and consumption by flora and fauna, it is necessary to discuss some of the environmental effects of light, temperature, CO_2, O_2 and nutrients that influence photosynthetic processes. But, we must understand that at the outset of this discussion an amazing apparent equilibrium exists between global primary production and global respiration. Why such an equilibrium exists will not be discussed here, but a surprising fact it certainly is since, fungi and bacteria (the largest contributors to respiration), which decompose organic matter (also living material), do so only to a certain extent, thus allowing the existing dynamic equilibrium to be maintained. Our problem is how a disequilibrium can occur and thus cause changes in atmospheric and oceanic oxygen.

Photosynthesis is a coordinated serial oxidation-reduction process, which includes a number of photochemical and enzymatic reactions. Assimilation of CO_2 occurs along either the Calvin-Benson Cycle (C_3 pathway), or the Hatch-Slack-Kortschak (C_4 pathway). In the C_3 pathway, the first produced organic compound, PGA (Phosphoglyceric acid) contains three carbon atoms. The primary enzyme involved is Rubisco (Ribulose-bi-PO_4 Carboxylase/Oxygenase), to which CO_2 and O_2 compete for photosynthesis or photorespiration respectively (Fig. 9.8).

In the C_4 pathway, there exists a primary and secondary CO_2 fixation cycle, the first occurring by the enzyme called PEPC (Phospho-enol-pyruvate-Carboxylase) which has a much higher affinity for CO_2 than Rubisco. The primary cycle is followed by a C_3 one in which the reduction to carbohydrate takes place. The name C_4 pathway is given because the first product is oxalo-acetic acid or OAA, containing 4 carbon atoms.

It are the differences between C_3 and C_4 plants with respect to external conditions such as light, temperature, moisture, nitrate, CO_2 and O_2 changes, that make them of major interest for potential atmospheric oxygen (and CO_2) regulation.

Various terms of interest are summarized:

Photosynthesis/photorespiration: C_3 plants can at the same time photosynthesize and photorespire (between 15 and 30% of photosynthesis). C_4 plants can only photosynthesize while photorespiration is not taking place (Edwards and Walker 1983).

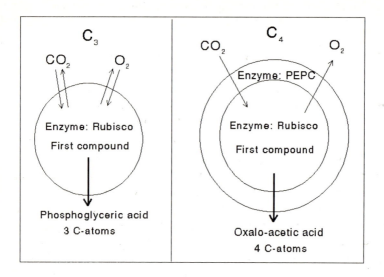

Fig. 9.8. Schematic view of C_3 and C_4 pathways of photosynthesis and respiration.

The different reactivity of C_3 and C_4 plants has also been modelled for the biosphere of the next 0.1 to 1.5 Gyr (10^9 year), when it is expected that due to diminishing atmospheric CO_2 levels C_4 plants can survive for a much longer period than C_3 plants (Caldeira and Kasting 1992).

Compensation concentrations: The CO_2 compensation concentrations are much lower for C_4 plants: 50 ppmv CO_2 (0.005 CO_2 vol%) for the C_3 enzyme Rubisco and 2 ppmv CO_2 (0.0002 vol% CO_2) for the C_4 enzyme PEPC (Black et al. 1969; Goudriaan and Ajtay 1979; Edwards and Walker 1983; Larcher 1983).

Depression by O_2: Depression of photosynthesis by O_2 occurs only in C_3 plants, not in C_4 plants. This is due to the competition of O_2 and CO_2 for Rubisco which either acts as oxygenase or carboxylase (Edwards and Walker 1983).

Nutrients: An essential difference exists between the nutrient demand of C_3 and C_4 plants (Schmitt and Edwards 1981). C_4 plants have a lower nutrient demand for the same photosynthetic activity than C_3 plants. This is of considerable interest for terrestrial regions and ocean surface waters with low nutrient concentration. On a global scale the ratio of C_3/C_4 species can potentially be influenced by the influence which man has on the world vegetation. At present the majority of terrestrial species are C_3 species, although the C_4 biomass of herbs in hot arid areas may override that of C_3

species (Hofstra and Stienstra 1977; Werger and Ellis 1981). Agriculture is focused on culture of economically valuable species, and C_4 plants (maize and sugar cane are examples of such C_4 plants) give a higher production per unit fertilizer. The C_4 plant maize has, for example, an efficiency of 6.7 mg CO_2/mg N/hr, while that of the C_3 plant rice is 5.9 mg CO_2/mg N/hr, determined both at 20 °C (Schmitt and Edwards 1981).

Light curve: Light-dependency of photosynthesis has a maximum (saturation point) for C_3 plants, but not for C_4 plants (Ehleringer 1978; Larcher 1983; Sengbusch 1989). Evidently PEPC, even at the strongest light intensities, is capable of keeping pace with existing light irradiation.

Temperature: Temperature affects photosynthesis through enzymatic processes. The photochemical process is nearly independent of temperature, but enzymatic fixation of CO_2 increases with temperature, until a maximum value is reached. Temperature has no effect on photosynthesis at low light intensities, however the photosynthetic rate will increase at intense light conditions for C_3 species, due to the sensitivity of Rubisco to temperature. High temperature favours photorespiration (in C_3 plants) by selective decrease of carbon dioxide relative to oxygen. In marine phytoplankton, Harrison and Platt (1980) and Côté and Platt (1983) have shown that 50% of the photosynthesis variations were due to temperature changes. For land plants, high temperature may cause desiccation, to which C_4 plants are better protected than C_3 plants. Temperature also determines the distribution of C_4 plants and perhaps of phytoplankton. C_4 land plants have a clear preference for higher temperatures, and will not be found in regions with frost (Werger and Ellis 1981).

Distribution of C_3 and C_4 plants: The number of terrestrial C_4 species in northern Europe is small and clearly depends on the climate (Collins and Jones 1985). From north to south the percentage of C_4 (in number of species) increases from zero in northern Europe to 4.4% in the Azores.

C_4 algae: Although algae are not characterized as C_4 species (Badger 1985; Descolas-Gros and Fontugne 1990), a majority of algae contain a C_4-like photosynthetic pathway which deviates from C_3 ones in their CO_2 concentrating mechanism. This is linked to the extra ATP requirements for concentrating CO_2 from extremely low levels just outside the algal cell. Whether this occurs for all algae is not certain, but PEPC activities are measured in several species (Berkaloff et al. 1981). When algal and cyanobacterial cells are depleted of CO_2, they switch to the uptake of CO_2 + HCO_3^- within themselves, using PEPC, resulting in repression of photorespiration. Hence phytoplankton is possibly adapted to ocean systems in which HCO_3^- represents 90% of ΣCO_2 ($=CO_2$ + HCO_3^- + CO_3^{2-}) at a pH ranging from 7.8 to 8.2.

$\delta^{13}C$: There exists both for terrestrial and aquatic plants a difference in carbon-isotope ratio of organic carbon created by C_3 and C_4 photosynthetic pathways. C_3 photosynthesis results in $\delta^{13}C$ values between -20 and 40 ‰ and C_4 between -10 and 20 ‰ (Larcher 1983). Thus, the $\delta^{13}C$ of vegetal organic matter can indicate the ratio of C_3 to C_4 photosynthesis producing that matter. Indicative are the recent findings of $\delta^{13}C$ changes of soil humus for historical C_3/C_4 plant coverage ratio which changed from C_4 plants (dry area) to C_3 plants (present subtropical forest area) in the State of Paraná, Brazil (23 °S, 53 °W) in a period of less than 10,000 years (Pessenda et al. 1993).

9.3.2 Short time scale oxygen regulatory feedbacks

An evaluation of oxygen regulatory feedbacks on a short time scale in atmosphere and oceans must take into consideration at each moment that oxygen levels can *only* increase when photosynthesis is larger than respiration (C_3 plant photorespiration and animal and bacterial respiration). This will result in burial of organic matter or increases of biomass. Feedbacks are given for terrestrial and aquatic systems with the major external factors (I) oxygen, (II) CO_2, (III) nutrients, (IV) temperature and (V) light, that would be of impact for 'short-term' oxygen regulation and burial of organic matter (Table 9.2).

Table 9.2. Short-term oxygen regulation

External factors	Terrestrial/aquatic conditions	Result (*positive or negative feedbacks are in italic*)
I: O_2 Terrestrial	Atmospheric oxygen levels can only change slightly on a short time scale due to the fact that they are negatively correlated to CO_2 in the order of 0.03 CO_2 vol%. No photosynthetic system will be sensitive enough to react on such a small deviation from 20.946 O_2 vol%.	There will be *no feedback* based on atmospheric oxygen changes.
O_2 Aquatic	In aquatic systems, O_2 can easily attain 200% of the solubility saturation level during high primary production or decrease to zero when bacterial respiration is high.	O_2 *can affect* only C_3 photosynthesis, since C_3 photosynthesis can shift to photorespiration at high oxygen levels.

External factors	Terrestrial/aquatic conditions	Result (*positive or negative feedbacks are in italic*)
II: CO_2 Terrestrial	Higher atmospheric CO_2 will cause greenhouse effects in the atmosphere and temperature increases also (see IV, temperature). High atmospheric CO_2 concentration will increase photosynthesis both in C_4 and in C_3 plants. Some C_4 plants can close their stomata prior to an increase of photosynthesis in order to reduce loss of water, so the temporary effect is less than for C_3 plants. Low CO_2 reduces C_3 activity earlier than C_4 activity because of the difference of their compensation points.	For both C_3 and C_4 species, increased productivity acts *only as feedback* for increasing or decreasing CO_2 when it is accompanied by an *increase or decrease, respectively* of biomass and humus formation, since otherwise production is balanced by respiration.
CO_2 Aquatic	Increased atmospheric CO_2 will only slightly affect the CO_2 level in the oceans, since the oceans contain 55 times more CO_2 than the atmosphere.	Thus there will be *little or no feedback* based on increased atmospheric CO_2.
III: Nutrients Terrestrial	Nutrient deficiencies will occur with strong biomass growth.	This *may act as a feedback* to limit primary productivity, biomass and humus formation and thus oxygen production.
Nutrients Aquatic	Nutrient deficiencies usually occur during high primary productivity in spring, summer and autumn. The corresponding high temperatures at the same time will stimulate surface layer stratifications and subsequently restrict nutrient fluxes from deeper layers.	Thus nutrients *play a role in the feedback*.
IV: Temp. Terrestrial	Higher temperatures favour the presence of C_4 plants. This will affect the C_3/C_4 species distribution in favour of C_4 species, and thus increase the total efficiency of photosynthesis/amount of nutrients. Since these higher temperatures are caused by greenhouse effects at elevated CO_2 levels, more CO_2, relative to nutrients will be bound.	This process is a *feedback to reduce the greenhouse effect* but may favour more biomass and humus formation and a resulting oxygen production.

Table 9.2. continued

External factors	Terrestrial/aquatic conditions	Result (*positive or negative feedbacks are in italic*)
Temp. Aquatic	Higher temperatures of surface ocean waters will favour stratification (Oppo and Fairbanks 1990), and thus reduce the exchange of nutrients between deeper and surface waters. On the other hand, burial of organic matter will be increased, thus causing an excess of photosynthesis over respiration.	The net effect is an oxygen release to the atmosphere and an excess uptake of CO_2 by the oceans. This will increase burial of organic matter (causing *oxygen overproduction*) but also act as feedback to reduce greenhouse heating. Whether there is a change from C_3 to C_4 photosynthesis, is not yet clear, although the different $\delta^{13}C$ values of phytoplankton in the oceans indicate a correlation with temperature (Beardall et al. 1976; Descolas-Gros and Fontugne 1990).
V: Light Terrestrial	If gas exchange is not restricted by environmental factors such as water supply and temperature, net photosynthesis parallels light availability up to a - saturation level for C_3 plants while there are no restrictions for C_4 plant growth. Photoinhibition in C_3 plants at excess irradiation usually occurs temporarily during mid-day, and is rarely irreversible.	Cloud cover, also a feedback of an increase in the greenhouse effect, cause similar change in photosynthetic rates for both C_3 (below saturation level) and C_4 plants. High light conditions act as *feedback* for CO_2 reduction and oxygen supply, at least when there *is humus formation*.
Light Aquatic	Algal growth depends on light absorption in the water column, which usually results in a maximum photosynthetic activity at some distance below the water surface, above which there is photoinhibition and below a photoadaptation with depth decreasing radiation.	Light has an impact on the photosynthetic processes and has the *same feedback* possibilities as *on land*.

Table 9.3.A. O_2 and CO_2 in atmosphere and oceans. O_2 data: Duursma and Boisson (1994), Broecker (1970) and Budyko et al. (1987); CO_2: using the tables of Buch et al. (1932) and from Bolin et al. (1979). Earth data Bowden (1965). **B.** O_2 and CO_2 ratios. **C.** Primary productivity data from Broecker (1970) and Berger et al. (1989).

A	ATMOSPHERE Earth surface: 5.1×10^{14} m^2 Continent surface: 1.5×10^{14} m^2		OCEANS Ocean surface: 3.6×10^{14} m^2 Ocean volume: 1.37×10^{18} m^3 Av. Ocean depth: 3.8×10^3 m		ATMOSPHERE + OCEANS (Total)	
	O_2	CO_2-C (1965)	O_2	CO_2-C	O_2	CO_2-C
Total mass in mol	3.75×10^{19}	5.3×10^{16}	3.1×10^{17}	2.9×10^{18}	3.78×10^{19}	2.95×10^{18}
Total mass in g	1.2×10^{21}	6.4×10^{17}	9.8×10^{18}	3.5×10^{19}	1.21×10^{21}	3.56×10^{19}
mol/m^2	7.35×10^4	1.04×10^2	8.6×10^2	8.1×10^3		
g/m^2	2.3×10^6	1.25×10^3	2.8×10^4	9.7×10^4		
mol/m^3 (1 atm.)	9.35	1.34×10^{-2}	2.3×10^{-1}	2.1		
g/m^3 (1 atm.)	299	0.161	7.4	25.0		

B

Atmosphere: total mol O_2/total mol CO_2	= 700
Oceans: total mol O_2/total mol CO_2	= 0.10 = 1/10
Total mol O_2 atm./total mol O_2 oceans	= 120
Total mol CO_2 atm./total mol CO_2 oceans	= 1.8×10^{-2} or 1/55

C (*)

Average terrestrial prim. prod.: 12 mol Org-C/m^2/yr = 12 mol O_2/m^2/yr
Total terrestrial prim. prod.: 1.836×10^{15} mol Org.-C/yr = 1.84×10^{15} mol O_2/yr
Average oceanic prim. prod.: 6 mol Org.-C/m^2/yr = 6 or 6.9 mol O_2/m^2/yr
Total oceanic prim. prod.: 2.16×10^{15} mol Org.-C/yr = 2.16 or 2.48×10^{15} mol O_2/yr
Total earth prim. prod.: $(4.0-4.3) \times 10^{15}$ mol Org.-C/yr = $(4.0-4.3) \times 10^{15}$ mol O_2/yr
This leads to a mean turnover time for (atmospheric plus oceanic) oxygen of 8800-9450 yr (compare 2000 yr: Hall and Rao, 1987; 6000 yr: Holland, 1978). The mean turnover time for (atmospheric and oceanic) carbon dioxide is 690-740 yr.

(*) Prim. prod. = primary production

9.3.3 Global short-term feedback possibilities

Considering all effects of external factors on photosynthesis, photorespiration, and animal and bacterial respiration, we can conclude that global short-term regulation of atmospheric oxygen is based on the same principles as long-term regulation. The difference is that the speed of humus formation (and its burial) or of its erosion cause short-time processes to be negligible. Nevertheless, it is the *living material* that is at the basis of any change in atmospheric oxygen variation (except for fossil-fuel burning) and there still may exist short-term regulation processes superimposed on the long-term variations. The question is, how short can such a superimposed variation be?

9.3.3.1 Global time scales

We have demonstrated that all free and bound oxygen on earth has been produced by living vegetal material, whose net production from 3.2 billion years ago until present is 15 times the present atmospheric oxygen mass (Fig. 9.1). We have also seen that each atmospheric oxygen molecule passes on average once each 9000 yr (turnover time) through living material (Table 9.3).

If consequently, photosynthesis were to fail hypothetically for a time span of 0.9 year ($1/10^4$ of turnover time) and total respiration were to remain constant, this would only change the atmospheric oxygen by a factor $1/10^4$ of 20.946 vol%, which is 0.0021 vol%/yr. If photosynthesis were to exceed total respiration by for example 1%, causing an equivalent increase of biomass and buried humus, the atmospheric oxygen would increase by (0.01/9000)x20.946 vol% = 0.000023 vol%/yr.

9.3.3.2 Time scales for oceans

The world oceans have a relatively small oxygen content which is 1/120 (Table 9.3) of that of the atmospheric oxygen reserve. However, oceans have a very large interface with the atmosphere, which is 70% of the earth's surface and a primary productivity which is ranging from 30 to 300 g C/m^2/yr (Berger et al. 1989). In spite of the fact that the oceans only contain 0.22% of the world's biomass standing stock, their primary productivity is greater due to a much more rapid turnover of the oceanic biomass, viz., 54% of the world primary production (cf. Table 9.3 and Taube 1992). In principle the ocean primary productivity is flexible enough to react on changes in O_2. Such O_2 variations are thought to be controlled by nitrogen-nutrients levels over periods of 1000 to 10,000 years (McElroy 1983). Such nutrient levels vary with changes in oceanic circulation and upwelling, sea-level rise, erosion and stratification. Since the exchange of oxygen between atmosphere and oceans is rapid, as we will see later, the oceans can probably contribute substantially to the regulation of atmospheric oxygen, as was already mentioned (Boer 1986) due to higher burial of organic matter in warmer periods.

The question is, however, over which period of time can a measurable effect be observed? An example calculation will make clear that for a 0.1 vol% change of atmospheric oxygen this is still a long period. A gain of 0.1 vol% of atmospheric O_2 would involve a flux of $(7.25 \times 10^4$ mol $O_2/m^2)$ x $(1/0.70)$ x $(0.1/20.946) = 494$ mol O_2/m^2. The average annual primary production of the world oceans is 2.5×10^{15} mol Org.-C/yr (Berger et al. 1989) or 6.9 mol $O_2/m^2/yr$. This is 72 times lower than the above hypothetical gain of 494 mol O_2/m^2. Supposing that compensation can occur through a 1 or 0.1 % ocean primary production increase over respiration (and subsequent burial of organic matter), the atmospheric adjustment will require 7,200 or 72,000 yr, respectively. To replace all 20.946 vol% O_2 would take 1.5x10⁶ or 1.5x10⁷ years, respectively. Such a cycle could occur in the 1.8 billion years of photosynthetic oxygen formation 1,200 or 120 times, respectively. Although these numbers have only illustrative value, they show the range in which changes could be encountered.

It is necessary to consider the effect of increased photosynthesis over respiration on the burial rate of organic matter in the oceans. For the 1% and 0.1% increase of the above given production of 6.9 mol $O_2/m^2/yr$ this results in an increased burial of 0.069 or 0.0069 mol $O_2/m^2/yr$, respectively.

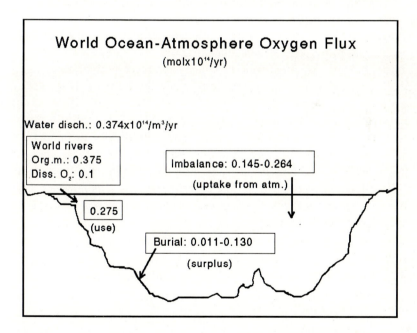

Fig. 9.9. World ocean-atmosphere oxygen fluxes (Duursma and Boisson 1994).

For the world oceans the actual average deposition rate of organic matter calculated from Fig. 9.9, is equivalent to 0.003-0.036 mol O_2/m^2/yr. Burial should therefore increase by a factor 2300 or 230%, respectively for the lowest burial value, and 190-19%, respectively, for the highest value. This is difficult to imagine, unless parts of the oceans become anoxic below the euphotic surface layer.

9.3.4 Final conclusion

The world terrestrial and oceanic systems of photosynthesis and respiration of flora and fauna contain the necessary components for regulation of atmospheric oxygen. The question as to how this actually occurs requires many detailed and global studies, but we can conclude that the relatively stable oxygen content of the atmosphere is the result of a complex short-term and long-term feedbacks. These feedbacks are primarily caused by the earth-sun interaction and resulting heat budget and cyclic world-climate changes, from which the glacial-interglacial cycles are the shortest. The feedback processes are able to maintain extremely constant oxygen concentrations over periods of thousands of years, superimposed on a multimillion cycle with extremes of oxygen concentration between 15 and 30 vol%.

No evidence is found for an autoregulation of atmospheric oxygen solely based on changes in oxygen. Oxygen concentrations are therefore the result of larger terrestrial and aquatic loops in which factors of temperature, light, nutrients and CO_2 play a role, and for longer periods elements like sulphur and iron may be involved. Hence the present level of 20.946 vol% of atmospheric oxygen is just a temporary one which will change in the course of millions of years. The question of an optimum concentration for sustaining life on earth is equally time-dependent, but bearing in mind that these changes occur in the order of millions of years, evolutionary processes will probably keep pace with the oxygen changes. The amazing fact remains however, that global photosynthesis equalizes so well global respiration, which is the keystone of the global stability of oxygen.

9.4 Future limit(s) of ocean pH due to fossil-fuel produced CO_2

An increase in CO_2 concentration slightly raises the acidity of sea water. The correlation at dynamic equilibrium between atmospheric CO_2 and ΣCO_2 in the oceans can probably be modelled for a hypothetical average ocean temperature, taking into account the increase of total (atmosphere + oceans)

CO_2, the effect of temperature on the two equilibrium constants K_1 and K_2 of the reaction (9.3), and the partial dissolution of carbonates. How this should be calculated is a specialized matter and is not treated in this overview. Nevertheless it is possible to estimate the limits of decrease the pH should undergo by way of plots, neglecting additional dissolution of carbonates.

Economically exploitable fossil fuel reserves are estimated at 1.8×10^{19} g of carbon (Keeling et al. 1993), which is, when burned, equivalent to 1.5×10^{18} mol CO_2. The present-day increase of atmospheric CO_2 mass is estimated at 2.3×10^{14} mol CO_2/yr (Fig. 9.7A), where the oceans receive 1.75×10^{14} mol CO_2/yr. The total contribution of these fluxes is 4.05×10^{14} mol CO_2/yr. At these rates, fossil-fuel will be exhausted within 3700 yr. By then, there will be added to the atmosphere $(2.3/4.05) \times 1.5 \times 10^{18}$ mol $= 8.5 \times 10^{17}$ mol CO_2, which is equivalent to 4811 ppmv. Hence the atmosphere will ultimately have a CO_2 concentration of 4811 + 300 (1965 value) = 5111 or about 5000 ppmv. With an annual increase CO_2 of 1.25 ppmv CO_2/yr (Keeling and Shertz 1992, Fig. 9.3) we come for 3700 years of continuous fossil-fuel CO_2 production to a slightly higher figure of 4925 + 300 ppmv = 5225 ppmv.

Assuming for the moment that the distribution ratio between atmosphere and oceans remains the same as given in Table 9.3, the total amount of additional CO_2 to be accumulated by the oceans will be $(1.75/4.05) \times 1.5 \times 10^{18} = 6.48 \times 10^{17}$ mol, before re-equilibration between atmosphere and oceans. This will increase the 1965 ocean CO_2 mass of 2.9×10^{18} mol to 3.55×10^{18} mol. The average concentration of ΣCO_2 of 2.1 mol/m^3 will thus increase to 2.57 mol/m^3 (ocean volume taken at 1.37×10^{18} m^3; Bowden 1965).

The assumption that atmospheric pCO_2 was in equilibrium with the dissolved ΣCO_2 in the pre-fossil fuel period is deducted from the temperature related pCO_2 as determined on the Antarctic Vostok ice core by Barnola et al. (1987) and plotted by Duursma and Boisson (1994). Partitioning between atmospheric and oceanic CO_2 in 1965 seems to approach this equilibrium too, as can be seen in Fig. 9.10.

In this figure, saturation curves are given for pCO_2 against ΣCO_2 in sea water (North Sea and Baltic) for 4 different temperatures and a salinity of 35 ‰, all based on the classical ICES tables determined and published by Buch et al. (1932). When we plot the ultimate atmospheric pCO_2 of about 5 mBar (5000 ppmv) and the corresponding ΣCO_2 ocean concentration of 2.57 mol/m^3, respectively, we can see that the dotted lines intersect at the saturation curve for 5 °C (Fig. 9.10). This is noteworthy, since atmospheric CO_2 uptake by the oceans is thought to be a retarded process due to the slow vertical mixing of water masses.

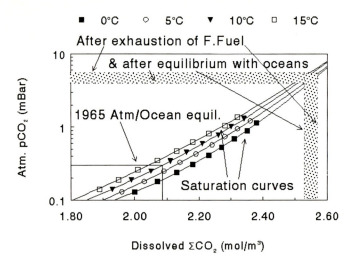

Fig. 9.10. Saturation curves of atmospheric pCO$_2$ and ΣCO$_2$ of sea water (North Sea and Baltic Sea; 35 ‰ salinity) for different temperatures as determined by Buch et al. (1932) at equilibrium pH. In this figure are plotted the 1965 figures on pCO$_2$ (0.3 mBar) and ΣCO$_2$ (2.1 mol/m^3) (Duursma and Boisson 1994), and those calculated (see text, page 221) for exhaustion of known exploitable fossil fuel reserves and the partitioning of CO$_2$ between atmosphere and oceans and subsequent equilibration.

But, is this an accurate assumption? The relatively large imbalance of (2.12-2.24)x10^{14} mol CO$_2$/yr with respect to the CO$_2$ flux from atmosphere to oceans (1.75x10^{14} mol CO$_2$/yr) suggests that we still lack precise information on the ocean CO$_2$ budget.

It can, however, be calculated accurately that for each 0.1 mBar reduction of atmospheric pCO$_2$ (equivalent to 1.77x10^{16} mol of total atmospheric CO$_2$), an increase of 0.013 mol/m^3 of oceanic ΣCO$_2$ is expected (1 mol/m^3 ΣCO$_2$ is equivalent to 1.37x10^{18} mol of total oceanic CO$_2$). On this basis a best fit can be calculated with final equilibrium values for pCO$_2$ lower than 5 mBar and ΣCO$_2$ just a fraction higher than 2.57 mol/m^3. This is shown in Fig. 9.10 for a pCO$_2$ reduction from 5 to 4 Mbar and a related ΣCO$_2$ increase from 2.57 to 2.69 mol/m^3.

The publication of Buch et al. (1932) provided tables on the correlation between pH, temperature and either pCO$_2$ or ΣCO$_2$ for different salinities. pCO$_2$ (in equilibrium with ΣCO$_2$) can be correlated with pH, whereas the influence of temperature (0 - 15 °C) is small (Fig. 9.11). Introducing the expected pCO$_2$ of 5 mBar (5000 ppmv) after fossil-fuel exhaustion, we arrive at a corresponding pH of 7.00 (Fig. 9.11).

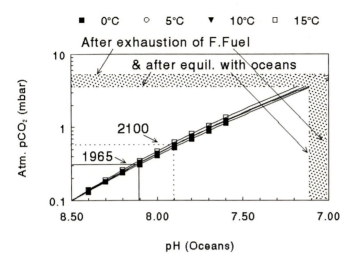

Fig. 9.11. Correlation curves of pCO_2 and pH in sea water (North Sea and Baltic Sea, S = 35 ‰) for different temperatures as determined by Buch et al. (1932) in equilibrium with ΣCO_2. In this figure are plotted the 1965 figures on pCO_2 (0.3 mBar) and pH (8.10) (See Duursma and Boisson 1994). The pCO_2 values of 5 and 4 mBar are taken from Fig. 9.10 representing the situation after exhaustion of known exploitable fossil fuel reserves and subsequent equilibration, respectively. The deduced pH's of 7.00 and 7.10 respectively, are obtained from intersection with the designed curves.

This will probably represent the pH of ocean surface waters in contact with the atmosphere. When later equilibrium between atmosphere and oceans is attained, and pCO_2 may decrease to 4 mbar, the final average pH of the oceans should be around 7.10. For example a potential pH of 7.90 can be expected when the atmospheric pCO_2 of 0.57 mBar (570 ppmv) is attained, which may happen in the year 2100 for an atmospheric increase of 2 ppmv CO_2/yr.

The reason why the old tables of Buch et al. (1932) were used was their completeness concerning all the parameters pCO_2, ΣCO_2, pH, temperature and salinity, measured during the same experiments and on natural sea water. More recent data did not have this combination of all five parameters.

In his chapter on CO_2, Skirrow (1965) compared Buch et al. (1932)'s values on the first and second apparent dissociation constants with those as determined by Lyman (1956). He found for pK_1' only a difference of ≤ 0.02 and for pK_2' a difference of ≤ 0.12, for temperatures from 0 to 30 °C, where pK_1' ranged from 6.16-5.99 and pK_2' from 9.42-9.00. These apparent dissociation constants are for the same salinity of 35 ‰ equal to those given by Wollast and Vanderborght (1994) and which range from 0-30 °C for pK_1': 6.19-

10.2 Techniques

10.2.1 Environmental Impact Assessment (EIA)

EIA is defined as the technique to evaluate the factors of value to the
environment with regard to proposed projects, if possible presented with
various industrial, economic or technical alternatives. The assessment should
produce a document in clear language, illustrated with schemes, presenting the
likely economical benefits and environmental damage involved. A distinction
is made between transient and permanent effects. These effects should be
given scores in terms of ecological values (e.g. rarity of species, diversity of
species, value to migrating species) or/and economical values (e.g. value of
ecosystem for agriculture or fisheries).

10.2.2 Best Professional Judgement (BPJ)

BPJ is a technique where a group of specialists of different disciplines testify
to their best level of competence on the various aspects on which an EIA is
based. The technique concerns:
- The selection of a number a specialists with regard to their knowledge of
  technological, economical and environmental factors.
- The selection of EIA factors and parameters that determine the
  technological, economical and environmental aspects.
- A 1^{rst} matrix: A scheme is prepared to arrive at an understandable
  assessment of the project proposal: horizontally the Project factors and
  vertically the EIA factors.
- Nomination of a rapporteur.
- The BPJ scores are expressed as: - - = double negative, for a permanent
  damage; - = negative, for a temporary damage; 0 = no effect; + = a
  beneficial effect, but temporary; and + + = a permanent beneficial effect.
- All BPJ these scores require separately a short explanation in an annex.
- A 2nd matrix is prepared in which the total scores Σ- - , Σ-, Σ0, Σ+, and Σ+ +
  are plotted, if possible with different colours, from red (- -) to green (+ +)
  for each of the technical and economical alternatives of the proposal.
- A 3rd matrix where the Σ- - are numbered starting with the lowest Σ- - to the
  highest Σ- -. In case of equal Σ- -'s the amount of Σ- is used for making the
  subsequent order.

- A final discussion, if necessary with a 4th matrix, evaluating all technical, ecological and economic score will be prepared, on which basis the responsible authorities can make decisions.

10.3 A Delta-case EIA study

In the Netherlands, after the large storm-surge inundations of 1953 (Duursma et al. 1982) it was decided to heighten the dikes and to construct dams and storm-surge barriers in the estuaries of the Rhine-Meuse-Scheldt delta (Fig. 10.1). The work started in 1956 (after a first reparation of the 100 dike breaks) and was completed in 1987.

In the tidal estuaries, dams were initially constructed by a system of (i) caisson closure and (ii) dropping stones from a cable-railway. A third method called sand closure (iii) was applied later by pumping and bulldozing sand (Fig. 10.2). The last method is cheaper than the first two ones, but requires slack tide during the final closure of the dam for at least a period of one day, which was not possible for the primary dams or barriers in the mouths of the estuaries.

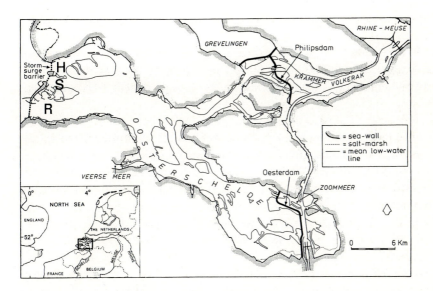

Fig. 10.1. Eastern Scheldt estuary in the SW Netherlands (insert). The three inlets of the storm-surge barrier are R = Roompot, S = Schaar and H = Hammen, by which the tides can be managed.

Cable-closure with stones

During tides

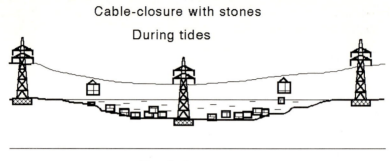

Sand-closure

Requires slack tide at final closure

Fig. 10.2. Techniques of dams construction.

In the Eastern Scheldt, where in 1985 a storm-surge barrier was completed, two secondary dams had still to be built. The construction of sand dams should cost 40 million US$ less than a stone dam, using cable-railway technique (Fig. 10.2).

The Ministry of Public Works (Bakker et al. 1984) requested EIA's for four alternative technologies of sand-dam constructions. Each of the technological alternatives was evaluated for all major seasons in which the work could be carried out. Dam construction were to make use of the storm-surge barrier in the mouth of the Eastern Scheldt (Fig. 10.1) for regulating of tides (normally 3.5 m amplitude).

The project technological alternatives were (Table 10.1) differed from each other in (i) amplitude and time of tide, and (ii) period of stagnancy: both arranged with the storm-surge barrier.

Fifteen environmental factors, including water quality, zoobenthos of tidal flats, salt-marsh vegetation, fishery, aquaculture etc., were classified by 6 scientists of different disciplines according to the scores -. - -. 0, + and + +. The percentage of - and - - are given in Table 10.2.

Table 10.1. Technical alternatives of tidal management with use of the storm-surge barrier.

Technical alternative	Tide	Tidal amplitude (cm)	Period (days)
I	T2	220	12
	stagnant	NAP	15
II	T2	220	12
	T1	220	20
	stagnant	NAP	3
III	T2	220	12
	T1	220	7
	T1	110	10
	stagnant	NAP	1
IV	T2	220	12
	T1	220	20
	T1	110	4
	stagnant	NAP	1

T2 = normal tide (2x/day); T1 = tide (1x/day); NAP = stagnant at national standard water level.

Table 10.2. Percentage of negative (-) and double negative (- -) scores as determined by BPJ.

Technical alternative	Scores	Spring	Summer	Autumn	Winter
I	-	97	95	95	91
	- -	79	88	79	61
II	-	48	63	38	24
	- -	4	19	4	0
III	-	51	56	49	35
	- -	15	15	5	0
IV	-	37	47	32	16
	- -	0	0	0	0

The scores of Table 10.2 were used to rank the technological alternatives, according to (i) the number of double negative scores and (ii) the number of single negative and (iii) the number of 0 scores, from 1 (best case) to 16 (worst case) and are given in Table 10.3.

Alternative IV/winter with the least change of tides and stagnancy, was environmentally the best technical alternative, followed by II/winter, IV/autumn and III/winter.

Table 10.3. Ranking of environmental impact assessments (EIAs) for the four different technical alternatives and for four seasons.

Technical alternatives	Spring	Summer	Autumn	Winter
	Ranking of EIAs by the %Σ- -, Σ- & Σ0, respectively			
I	15	16	14	13
II	8	15	7	2
III	10	11	9	4
IV	5	6	3	1

Each of the technical alternatives has a financial paragraph, being different for each of the four seasons. These have to be grouped into a separate table. Hence a second BPJ is to be made between the rapporteur of the EIA group and the project manager, together with technological and economic experts. This resulted in a recommendation to the policymakers, giving them the possibility to decide.

10.4 Assessment of diffusion of radionuclides from potential HLW disposal

In the 1980s an international study programme called SEABED   was completed to determine the potentiality of storage of high-level (nuclear) waste (HLW) in the deep ocean bottom. Such high-level nuclear waste requires several thousands of years of safe storage before the radioactivity will be reduced to permissable low levels (Fig. 10.3).

At present the number of nuclear power stations in operation and under construction is 432 and 48, respectively (Fig. 10.4). High-level nuclear waste amounts to 50,000 m^3 (OECD countries only) of 7×10^{16} Bq/m^3 (IAEA 1992), which is 3.5×10^{21} Bq. This suggests a global level of about 10^{22} Bq. If distributed equally over the oceans (volume 1.37×10^{18} m^3) this would result in concentrations of 73×10^3 Bq/m^3 or 2 μCi/m^3.

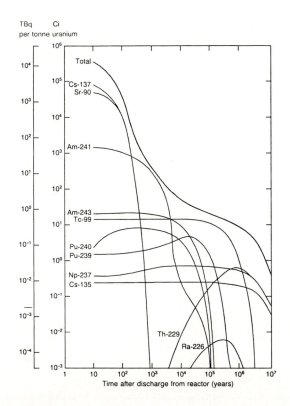

Fig. 10.3. Activity of high-level nuclear waste through time. Reprocessing is assumed ten years after discharge of spent fuel from reactor (reproduced from IAEA 1992).

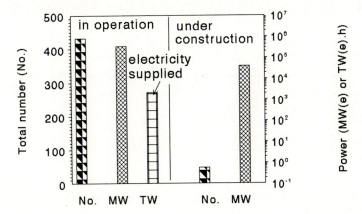

Fig. 10.4. Nuclear power reactors in operation and under construction, in number, their power and electricity output as of the end of 1994 (IAEA 1995).

In comparison, this is a factor 36 higher than the highest fallout levels found in the oceans (Duursma 1972) in the late sixties of 2×10^3 Bq/m³, indicating a matter of concern (see also Fig. 7.3; IAEA 1984; 1986b).

The principle idea of safe storage of nuclear waste, was to place canisters deep in the seabed for periods of at least ten to one hundred thousand years (Fig. 10.5). The studies concerned with addressing the need for safe long-term storage site qualification, engineering, safety assessment, legal and institutional aspects (NEA-OECD 1984) and the possible influence of temperature on the migration of radionuclides in deep-sea sediments (Geldermalsen and Wegereef 1985).

However, in the absence of accidents and breakup of storage canisters, storage may be regarded as safe, when deposited below 30 or 60 m of clay sediments for which the absorption properties of the clays would be known. (Duursma et al. 1983; Geldermalsen and Wegereef 1985). Releases from leaking canisters was calculated to be relatively small, even for nuclides with half-lives larger than 5000 years (Fig. 10.6).

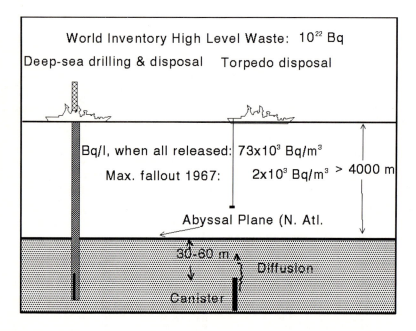

Fig. 10.5. Two scenario's for potential disposal of high-level nuclear waste into deep-sea abyssal plane sediments as investigated by the international programme SEABED. Left: by deep-sea drilling; right: by free-fall disposal.

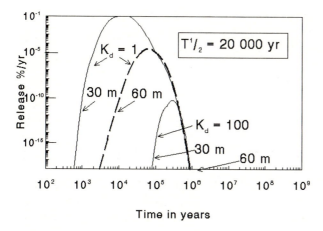

Fig. 10.6. Model calculation of annual % fraction of radionuclide waste as released from the seabed for two burial depths of 30 and 60 m and distribution coefficients (K_d) of 1 and 100, respectively. Usually the K_d's of the radioactive waste isotopes range between 10^3 and 10^5 (IAEA 1985), indicating that the annual fraction of release is smaller that 10^{-18} %/yr.

Table 10.4. Selected radioactive waste disposal. The 1990 stock of high-level nuclear waste was about 10^{22} Bq from which the OECD has 3.5×10^{21} Bq (50,000 $m^3 \times 7 \times 10^{16}$ Bq/m^3; IAEA 1992).

	Σ alpha	Σ bèta & gamma	References
Sellafield effluent	3.5×10^{13} (1980; Pu+Am)	5.53×10^{15} Bq (1980; Ru, Sr, Ce, Zr, Cs, H)	Pentreath (1985)
La Hague effluent	6.40×10^{11} Bq (1982)	1.26×10^{15} Bq (1982)	Calmet and Guegueniat (1985)
NE Atlantic solid waste	6.6×10^{14} Bq (1949-1982)	4.0×10^{16} Bq (1949-1982)	OECD-NEA (1985)

The reason why the SEABED programme did not proceed toward actual disposal of high-level radioactive waste was political. Public protests had already halted the ENEA disposal of low-level radioactive waste (LLW) in concrete containers on the seafloor, despite the fact that the total amount discharged over many years is of a similar magnitude (Table 10.4) as that annually released in liquid form into the Irish Sea and the English Channel from reprocessing plants. These results show that management of wastes and the use of the oceans for waste disposal cannot be based on scientific arguments only. The public, governmental and regulatory agencies must all be involved in environmental policy determinations such as high level waste disposal practices.

10.5 Amoeba assessment

In a simpler way, domestic, industrial and environmental parameters can be qualitatively assessed in a so-called amoeba configuration (Fig. 10.7), which presents a view of the competing factors and parameters that play a role in environmental management. This amoeba configuration was first introduced for description and assessment of ecosystems by the Dutch Ministry of Transport and Public Works (Brink et al. 1991) as a concept for sustainable economical development and environmental protection.

Such an assessment facilitates discussions on short and long-term approaches to environmental management and helps to facilitate a coordinated effort among different authorities at various national and regional levels and specialists having domestic, industrial and environmental responsibilities or interest.

The amoeba configuration represents each to domestic, industrial and environmental management factor by a segment of a circle (Fig. 10.7), where the line of the circle represents an apparent 'normal' situation. For a case in which local governments choose for increased tourism (No. 1 in Fig. 10.7), shown by an enlargement of the segment outside of the 'normal' circle, the consequences of such a decision can be estimated. In the example given, increased tourism and leisure (1) will be favourable for aquaculture and the environment (2), favour air quality (3) and employment (4), is negative or competitive for space with industrial development (5), will stimulate harbour development (marines) (6) and is positive for the construction of second homes and hotels (7). The local authorities will be faced with negative stress on the infrastructures (8), roads (9), sewage treatment (10) and tap water supply (11), which will require increased investments. From the resulting analysis in Fig. 10.7, large-scale fishery development does not match with tourism development (harbour, fishery industry) (12).

Once such an amoeba representation has some solid basis, the various costs and benefits have to be capitalized, both for the private sector and for the regional authorities. Any of the factors may be further evaluated re-applying the methodologies described above.

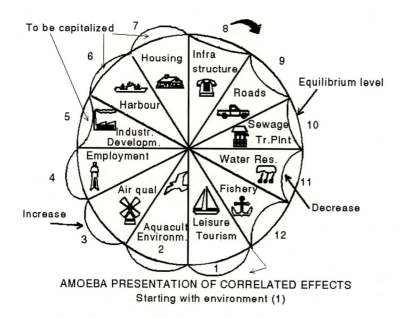

AMOEBA PRESENTATION OF CORRELATED EFFECTS
Starting with environment (1)

Fig. 10.7. Amoeba presentation of EIA factors of coastal-zone management.

10.6 COSMO-BIO assessment

It has been demonstrated at the 1993 conference on coastal zone management in Noordwijk (Netherlands; Bijlsma 1994) that only integrated management would lead to fruitful results for sustainable coastal development and protection. When industrial and urban development do not take into account negative long-term effects, the initial profits will be later counteracted by serious problems of environmental contamination, loss of prosperity and unemployment.

The Noordwijk conference has shown that for sustainable development to be achieved, many aspects of the human-environmental relationship must be considered, e.g. industrial activities, unemployment, gross product, housing, hotels, roads, harbour construction, fisheries, fresh-water sources, sewage treatment, infrastructure and leisure (Rijsberman et al. 1995). This is visualized in the attached disk COSMO-BIO, which presents a Demo of a coastal zone management situation for a temperate coastal environment. COSMO-BIO shows the impact of alternative user-specified coastal zone management for a synthetic case study area, 'Whale Bay', on a set of socio-economic criteria and 'population risk values' for a number of coastal organisms, characterised by biodiversity criteria.

The socio-economic and environmental criteria concern: fisheries, shipping, harbour activities, sand extraction, tourism, industry, water quality and nature conservation and biodiversity of the coastal environment. Parameters used are employment, standard of living related to the GRP (gross regional product), tourism development, cost of environmental measures, including limitation of pollution.

This model incorporates the methodology of research in the Netherlands to assess the risks of multiple stress caused by various human activities for a hypothetical coastal area in the temperate zone (Fig. 10.8). See further Appendix III, COSMO-BIO and the attached diskette.

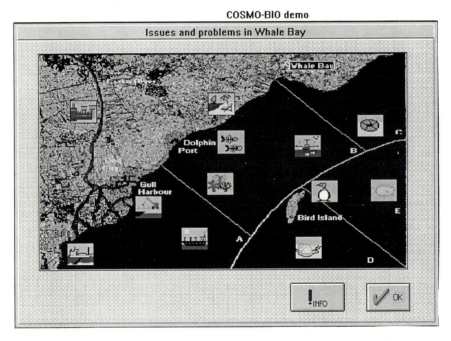

Fig. 10.8. Hypothetical coastal zone for which the role of biodiversity in coastal management is illustrated in the attached COSMO-BIO demo.

Appendix I. Answers to a question and exercises

Question preface.

The question of Prof. Kalle suggests that DOC below the euphotic zone originates from the primary production of that zone. However, deep oceans waters originate from polar regions where during winter downwelling occurs of surface water to large depths with subsequent horizontal transport to lower latitudes. These sinking polar waters have in winter low DOC contents which differ little from those found in the deep ocean water masses at great distance. Hence these deep DOC contents at lower latitude should not be compared to those of the surface layers. Any significant contribution through settling POC is a matter of much longer period than years. An equilibrium might be attained between this contribution and decomposition, before speaking of refractory DOC (Duursma 1961; 1965).

Exercise 2.1.

A principal non-causality should exist between the parameters presented under A, C and D. For A: D_{50} is determined on the basis of the % weight distribution of *all* grain sizes, and not just those of <63 μm and <25 μm. In absence of these fractions there is still a value for D_{50}, while in presence of these fractions no information is available of the weight percentage of the other fractions. The same argument is valid for the correlations C and D. Nevertheless the comparison of sorption on marine sediments is very often only correlated to either the clay-minerals or silt (smallest fractions), supposing that these fractions act as major sorption sites for contaminants.

Exercise 2.2.

$[Cu_{diss}] = 0.4xQ$ μg Cu/l water, which leaves for $[Cu_{part}] = 0.6xQ$ Cu μg/20 mg PM. K_d can be calculated as follows:

$$K_d = \frac{0.6 \cdot Q \ \mu g \ Cu/0.020 \ g \ PM}{0.4 \cdot Q \ \mu g \ Cu/1000 \ g \ water} = 7.5 \cdot 10^4$$

Check this number by applying formula 2.18.

Exercise 2.3.

Suppose the total amount of ^{239}Pu, particulate + dissolved) in one litre is Q Becquerel, from which X % is dissolved and (100-X) % is particulate. Then the K_d can be described as:

Check this percentage by applying formula 2.18.

$$K_d = \frac{(100-X)\cdot Q \ Bq/10\cdot10^{-6} \ gPM}{X\cdot Q \ Bq/1000 \ g \ water} = 5\cdot10^4$$

$$50\cdot X = 10^7 - 10^5\cdot X$$

$$X = 99.95 \ \%$$

Exercise 3.1.
The nuclide concentrations at equilibrium (Fig. 3.6) are:
(i) The PM suspension contains 10 Bq/ml,
(ii) the plankton suspension contains 11 Bq/ml, while (iii) there is 5 Bq/ml in solution. The suspensions contain 0.1 mg PM/ml and 0.50 μg plankton/ml (both dry weight).
   The nuclide concentrations for 1 ml suspension are: (10-5=5) Bq/0.1 mg PM and (11-5=6) Bq/0.05 μg plankton, respectively.
   Thus: K_d (PM) = {(5/0.1)x1000}:5 = 10^4 [(Bq/g):(Bq/g)], and K_d (pl) = {6/0.5)x10$_6$}:5 = 2.4x10^6 [(Bq/g):(Bq/g)], taking 1 ml water as 1 g water.

Exercise 3.2.
K_d = (150-120)/(10/1000) = 3000 [(μg/g):μg/g)], taking 1 l = 1000 g.

Exercise 3.3.
The amount of Bq/mol = -dN/dt = λN or 0.693N/T$_{1/2}$, N being the number of Avogadro, 6.02x10^{23}. For 1 mol ^{237}Pu this amounts to [0.693x6.02x10^{23}]/ [45x(24x3600)] or 1.073x10^{17} Bq/mol (T$_{1/2}$ in seconds, since Bq is number of disintegrations/second).
   The answer as requested per gram is: 4.53x10^{14} Bq/g.
   The same calculation can be made for ^{239}Pu with T$_{1/2}$ = 7.695x10^{11} seconds: 0.693x6.02x10^{23}/7.695x10^{11}x239 = 2.27x10^9 Bq/g.
   The answers in units of curies (1 Ci = 3.7x10^{10} Bq) are for ^{237}Pu = 1.224x10^4 Ci/g and for ^{239}Pu = 0.061 Ci/g.
   The general formula is for Bq/g: 4.17x10^{23}/(T$_{1/2}$xA), where A = atomic weight.

Exercise 5.1.
Suppose X is the PCB concentration in water and Y in air. The conversion of the PCB concentration ng/dry weight to μg/kg lipid results in: 300/0.03 = 10^4 μg congener/kg lipid.
   K_d (lipid/water) is given as 0.5x10^6 = (10^4 μg/kg)/(X μg/kg water), the value of X = 0.02 μg/kg or 20 μg/m^3 water.

K_d (water-air) is given as 122 (m^3/m^3). Its reciprocal value, K_d (air-water), is thus 8.2x10^{-3} (m^3/m^3) or (Y μg/m^3 air):(20 μg/m^3 water). The value Y becomes: 4.1x10^{-4} μg/m^3 or 410 pg/m^3.

The result is at the order of magnitute as given in Fig. 7.20B.

Exercise 7.1.

Inflow Gibraltar = 53,000 km^3/yr; Outflow Gibraltar = 50,500 km^3/yr. 1 nmol Cd = 112.41 ng.

Inflow in μg and ton is: 53,000x10^9 m^3/yr x 0.027x112.41 μg/m^3 = 1.61x10^{14} μg/yr = 161 ton/yr.

Outflow in μg and ton is: 50,500x10^9 m^3/yr x 0.078x112.41 μg/m^3 = 4.43x10^{14} μg/yr = 443 ton/yr.

Exercise 7.2.

The bottom layer concerned (10 cm top layer) has a volume of 3.6x10^{14} m^2 x 0.01 m = 3.6x10^{12} m^3. Taking for wet sediment, containing 40% pore water and a specific weight of 1.7 g/ml, this results in 0.6x6.12x10^{18} = 3.67x10^{18} g dry sediment.

The concentration of PCBs in the sediment can be calculated from the K_d (=10^6) and the PCB concentration in the water (17x10^{-12}/10^3 g water) = 1.7x10^{-8} gPCB/g sediment.

The total amount of PCBs in a 10 cm bottom layer is: 3.67x10^{18} x 1.7x10^{-8} = 6.24x10^{10} g PCB or 6.24x10^4 ton PCB.

Exercise 8.1.

$$PC_s = 1 - PC_w$$

$$PC_s = 1 - \frac{1}{1 + K_dS}$$

$$PC_s = \frac{1 + K_dS - 1}{1 + K_dS} = \frac{K_dS}{1 + K_dS}$$

Appendix II. Kara Sea box model user information

II.1. Summary

The Kara Sea box model (KASE) is a numerical code designed to predict radionuclide concentrations with time in four regions of the Kara Sea. The regions are Abrosimov Bay, the Novaya Zemlya Trough, the Eastern Kara Sea and the Western Kara Sea. Changes in concentration result from continuous or instantaneous releases of radionuclides from radioactive dumping grounds located in Abrosimov Bay, the Novaya Zemlya Trough, or the Ob and Yenisey Rivers and from the transfer of radioactivity among regions by water mass exchange (Fig. II.1). This documentation includes information on the numerical implementation of the code, an example simulation and the code listings.

II.2. System requirements

CPU: 80386 minimum, recommend 80486
Coprocessor: recommended
Hard Drive: recommended with 2-5 Megabyte available.
Floppy: 1.2 or 1.44 M floppy
Monitor: CGA, EGA, VGA, or Monochrome
Printer/Plotter: not supported
Mouse: not supported
MS-DOS: Version 5.0 or higher.

II.3. Code design

The KASE Box Model is configured to run on an IBM compatible computer (e.g. DOS platform) in base memory. The KASE Model consists of three main components: the executable file (KASE.EXE) and a series of input and output files (Table II.1). The executable program was coded and compiled using Microsoft FORTRAN Version 5.1.

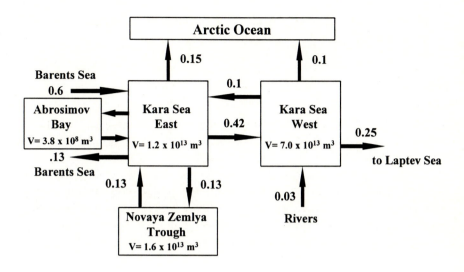

Fig. II.1. Structure of the Kara Sea box model (KASE). Water fluxes in units of Sverdrups occur in the direction of the arrows (after Sazykina and Kryshev 1994).

Table II.1. Files needed to execute the KASE Box Model.

Executable	Input	Output
KASE.EXE	ACTIV.DAT	KASE.OUT
	FLUX.DAT	COMMA.OUT
	BATCH.DAT	SPACE.OUT

There are two file types that are either used or created by the KASE model: *.DAT and *.OUT. These file types are in ASCII format so they can be viewed and edited in a DOS based text editor (i.e. EDIT in DOS Version 5.0). The symbol '*' is used as a "wildcard character" to delineate filenames with the same extension.

II.3.1. Executable file

The foundation of the KASE Box Model is the executable file, KASE.EXE (Fig. II.2). This program directs the operation of the numerical algorithm, accesses the input files and stores the results in the output files. The file

KASE.EXE is the compiled version of the source code, KASE.FOR (not provided) and cannot be accessed using the DOS editor.

II.3.2 Input files

Most of the preliminary information required for program execution is requested from the user after start-up through a series of screen prompts (see Figs. II.3 and II.4).

Note that all numeric responses to screen prompts should be given in the form of a real number except for responses to the number designation of the radionuclide of interest and the total time the model runs. All alphabetic responses to screen prompts must be given in capital letters. Failure to enter real numbers and capital letters at user prompts and in the BATCH.DAT file will result in errors during program execution.

REAL	INTEGER
3.0	3
3.	3
7.5	-

Two input files supply the model with additional initial information needed to successfully execute the KASE.EXE program. The input file ACTIV.DAT contains data on the background activities for the radionuclides used in the model for each region of the model (Fig. II.5). The FLUX.DAT file contains data on the rate of water mass exchange (flux) in and out of each of the boxes in the model (Fig. II.6). A user need not access these files because the files already contain the appropriate data. The program will automatically retrieve the information in these files during program execution. The user may wish to conduct additional experiments by changing the data in these files once he/she is familiar with the program. If this is the case, the data in ACTIV.DAT and FLUX.DAT files may be accessed and altered using the DOS editor.

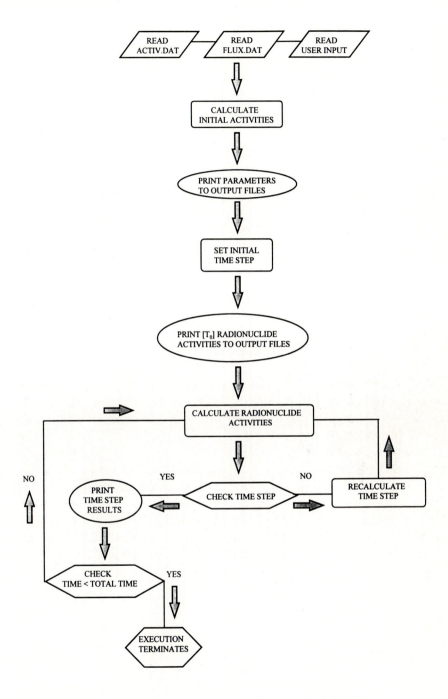

Fig II.2. Flowchart of program logic steps during execution of KASE.EXE.

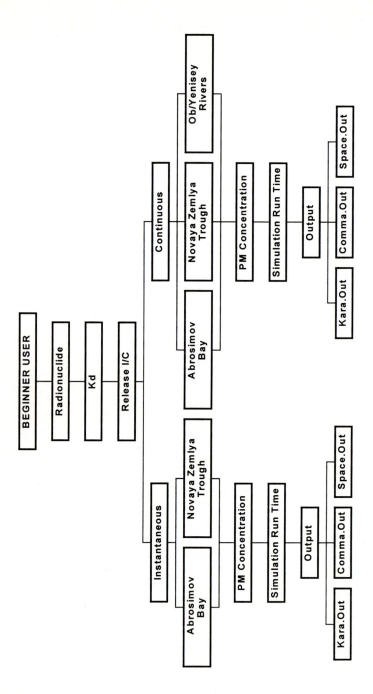

Fig. II.3. Series of screen prompts for a beginner user.

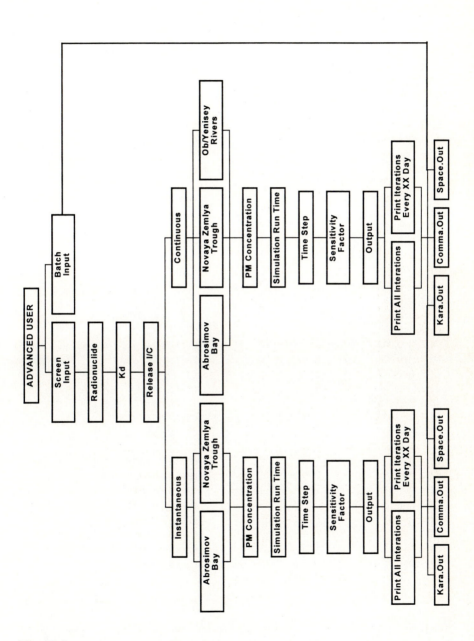

Fig. II.4. Series of screen prompts for an advanced user.

```
C BOX MODEL INPUT PARAMETERS
C INITIAL CONCENTRATION OF RADIONUCLIDES IN EACH BOX
C
C     1=PLUTONIUM-238              C     5=CESIUM-137
C1(1,1)=0.0003                     C1(5,1)=6.0
C1(1,2)=0.0003                     C1(5,2)=6.0
C1(1,3)=0.0003                     C1(5,3)=6.0
C1(1,4)=0.0003                     C1(5,4)=6.0
C1(1,5)=0.0003                     C1(5,5)=6.0
C1(1,6)=0.0003                     C1(5,6)=6.0
C1(1,7)=0.0003                     C1(5,7)=6.0
C1(1,8)=0.0003                     C1(5,8)=6.0
C     2=PLUTONIUM-239              C     6=AMERICIUM-241
C1(2,1)=0.003                      C1(6,1)=0.0006
C1(2,2)=0.003                      C1(6,2)=0.0006
C1(2,3)=0.003                      C1(6,3)=0.0006
C1(2,4)=0.003                      C1(6,4)=0.0006
C1(2,5)=0.003                      C1(6,5)=0.0006
C1(2,6)=0.003                      C1(6,6)=0.0006
C1(2,7)=0.003                      C1(6,7)=0.0006
C1(2,8)=0.003                      C1(6,8)=0.0006
C     3=PLUTONIUM-240              C     7=STRONTIUM-90
C1(3,1)=0.003                      C1(7,1)=5.0
C1(3,2)=0.003                      C1(7,2)=5.0
C1(3,3)=0.003                      C1(7,3)=5.0
C1(3,4)=0.003                      C1(7,4)=5.0
C1(3,5)=0.003                      C1(7,5)=5.0
C1(3,6)=0.003                      C1(7,6)=5.0
C1(3,7)=0.003                      C1(7,7)=5.0
C1(3,8)=0.003                      C1(7,8)=5.0
C     4=CESIUM-134                 C     8=COBALT-60
C1(4,1)=0.15                       C1(8,1)=0.0002
C1(4,2)=0.15                       C1(8,2)=0.0002
C1(4,3)=0.15                       C1(8,3)=0.0002
C1(4,4)=0.15                       C1(8,4)=0.0002
C1(4,5)=0.15                       C1(8,5)=0.0002
C1(4,6)=0.15                       C1(8,6)=0.0002
C1(4,7)=0.15                       C1(8,7)=0.0002
C1(4,8)=0.15                       C1(8,8)=0.0002
```

Fig II.5. Input file ACTIV.DAT stores background radionuclide activities for regions of the Kara Sea. The file is accessed by the program KASE.EXE during program execution.

```
C BOX MODEL INPUT PARAMETERS
C FLUXES (IN SVERDRUPS 10^6 m^3/s)
C
C FROM ABROSIMOV BAY                   C FROM OB/YENESEY RIVERS
F(1,1)=0.0                             F(5,1)=0.0
F(1,2)=0.0                             F(5,2)=0.0
F(1,3)=0.00005                         F(5,3)=0.0
F(1,4)=0.0                             F(5,4)=0.03
F(1,5)=0.0                             F(5,5)=0.0
F(1,6)=0.0                             F(5,6)=0.0
F(1,7)=0.0                             F(5,7)=0.0
F(1,8)=0.0                             F(5,8)=0.0
C                                      C
C FROM NOVOYA ZEMLYA TROUGH            C FROM LAPTEV SEA
F(2,1)=0.0                             F(6,1)=0.0
F(2,2)=0.0                             F(6,2)=0.0
F(2,3)=0.13                            F(6,3)=0.0
F(2,4)=0.0                             F(6,4)=0.0
F(2,5)=0.0                             F(6,5)=0.0
F(2,6)=0.0                             F(6,6)=0.0
F(2,7)=0.0                             F(6,7)=0.0
F(2,8)=0.0                             F(6,8)=0.0
C                                      C
C FROM WESTERN KARA SEA                C FROM BARENTS SEA
F(3,1)=0.00005                         F(7,1)=0.0
F(3,2)=0.13                            F(7,2)=0.0
F(3,3)=0.00                            F(7,3)=0.6
F(3,4)=0.42                            F(7,4)=0.0
F(3,5)=0.0                             F(7,5)=0.0
F(3,6)=0.0                             F(7,6)=0.0
F(3,7)=0.13                            F(7,7)=0.0
F(3,8)=0.15                            F(7,8)=0.0
C                                      C
C FROM EASTERN KARA SEA                C FROM ARCTIC OCEAN
F(4,1)=0.0                             F(8,1)=0.0
F(4,2)=0.0                             F(8,2)=0.0
F(4,3)=0.1                             F(8,3)=0.0
F(4,4)=0.0                             F(8,4)=0.0
F(4,5)=0.0                             F(8,5)=0.0
F(4,6)=0.25                            F(8,6)=0.0
F(4,7)=0.0                             F(8,7)=0.0
F(4,8)=0.1                             F(8,8)=0.0
```

Fig II.6. Input file FLUX.DAT stores water fluxes between regions of the Kara Sea. The file is accessed by the program KASE.EXE during program execution.

The BATCH.DAT input file is available for users familiar with the DOS editor and who prefer to bypass most of the screen prompts during execution of the program. If directed by the user at the appropriate screen prompt, the program will automatically retrieve input information from the batch file rather than from the screen. This option is available if the user chooses to conduct the current program execution as an Advanced User. The BATCH.DAT file provided on the program disk contains a sample input file that is fully operational in its present form and needs no additional editing before execution (Fig. II.7).

II.3.3 Output files

Output files contain information from the most recent model execution. Three output files provide alternative data formatting. The header information in each output file contains the date and time of the model execution (retrieved from the computer timeclock) and a summary of the input information provided by the user through screen prompts or through the BATCH.DAT file. Following the header information are the resulting model calculations of radionuclide activities (Bq/m^3) for Abrosimov Bay, the Novaya Zemlya Trough and the Eastern and Western Kara Sea. Calculations are stored after each time step during model execution. The output file KASE.OUT presents the results of the previous model execution in a form that is easily understood. The user may peruse the output information and make a general assessment of the results before deciding how to conduct follow-on executions of the model.

The output files COMMA.OUT and SPACE.OUT are designed to provide the user with files that may be imported into a spreadsheet for graphical analysis of temporal changes in radionuclide activities in each of the boxes (Section II.5). The two files are identical in format however COMMA.OUT records commas between data on a single line and SPACE.OUT records blanks in place of the commas. Before importing these files to a spreadsheet or graphics program it may be necessary to remove the header information from the files using the DOS editor. The format of the various output files may be seen in Figs. II.8 and II.9 and II.10.

```
//////////////////////////////////////////////////////////////////
BATCH FILE: PROGRAM KASE.EXE CONTAINING SIMULATION INPUT INFORMATION///
//////////////////////////////////////////////////////////////////
2      /RADIONUCLIDE/
10.     /DISTRIBUTION COEFFICIENT in KG/M^3 (.1-20000)/
I      /RELEASE INSTANTANEOUS OR CONTINUOUS/
1.      /ABROSIMOV BAY RELEASE AMOUNT in TBq = 10^12 Bq (1-1000)/
25.     /ABROSIMOV BAY PM CONCENTRATION (MG/L) (1-500)/
1.      /NOVAYA ZEMLYA TROUGH RELEASE AMOUNT in TBq = 10^12 Bq (1-1000)/
5.      /NOVAYA ZEMLYA TROUGH PM CONCENTRATION (MG/L) (1-10)/
0.      /OB/YENISEY RIVERS RELEASE AMOUNT in TBq/yr = 10^12 Bq/yr (1-10)/
50.     /OB/YENISEY RIVERS PM CONCENTRATION (MG/L) (1-500)/
18250   /SIMULATION RUN TIME IN DAYS (<182500)/
365.    /SIMULATION TIME STEP; DEFAULT =0.0/
2.      /SENSITIVITY FACTOR 1; DEFAULT =2.0/
2      /PRINTOUT 1=ALL OR 2=EVERY XX DAY/
1825    /EVERY XX DAY/
//////////////////////////////////////////////////////////////////
```

Fig II.7. Advanced users may use the input file BATCH.DAT to store the information needed by the program KASE.EXE during program execution.

DATE = 01/18/1996
  TIME = 09:44

              OUTPUT FOR KARA SEA BOX MODEL

     INPUT PARAMETERS

     SIMULATION RUN TIME -   18250 DAYS
     INSTANTANEOUS RELEASE
     RADIONUCLIDE = PLUTONIUM-239
     HALF-LIFE: .2411E+05 YRS
     DISTRIBUTION COEFFICIENT (Kd) - .10E+02 m^3/kg
     SENSITIVITY FACTOR -   2.00

  RELEASE LOCATION - ABROSIMOV BAY
       RELEASE AMOUNT -   1.00 TBq
       PM CONCENTRATION - 25.00 mg/liter
       PARTITION COEFFICIENT -   .80

  RELEASE LOCATION - NOVAYA ZEMLYA TROUGH
       RELEASE AMOUNT -   1.00 TBq
       PM CONCENTRATION -   5.00 mg/liter
       PARTITION COEFFICIENT -   .95

     LOCATION   RADIONUCLIDE ACTIVITY (Bq/m^3)
  XXXXXXXXXXXXXXXXXXXXXXXXXXXXXXXXXXXXXXXXXXXXXXXXXXXX
       TIMESTEP -   .00 DAYS
       ABROSIMOV BAY   .210526615E+04
       NOVAYA ZEMLYA TROUGH   .625238095E-01
       WESTERN KARA SEA   .300000000E-02
       EASTERN KARA SEA   .300000000E-02
  XXXXXXXXXXXXXXXXXXXXXXXXXXXXXXXXXXXXXXXXXXXXXXXXXXXX
       TIMESTEP - 1825.00 DAYS
       ABROSIMOV BAY   .923950618E-02
       NOVAYA ZEMLYA TROUGH   .320821888E-01
       WESTERN KARA SEA   .884456897E-02
       EASTERN KARA SEA   .957435159E-02
  XXXXXXXXXXXXXXXXXXXXXXXXXXXXXXXXXXXXXXXXXXXXXXXXXXXX
       TIMESTEP - 18250.00 DAYS
       ABROSIMOV BAY   .300345117E-02
       NOVAYA ZEMLYA TROUGH   .300910622E-02
       WESTERN KARA SEA   .300331834E-02
       EASTERN KARA SEA   .301368781E-02
  XXXXXXXXXXXXXXXXXXXXXXXXXXXXXXXXXXXXXXXXXXXXXXXXXXXX

Fig II.8. Example output file KASE.OUT. The file contains summary information for the previous program execution. The file is accessed for viewing using the DOS editor.

DATE = 01/18/1996
  TIME = 09:44

       OUTPUT FOR KARA SEA BOX MODEL

    INPUT PARAMETERS

      SIMULATION RUN TIME -   18250 DAYS
      INSTANTANEOUS RELEASE
      HALF-LIFE: .2411E+05 YRS
      DISTRIBUTION COEFFICIENT (Kd) - .10E+02 m^3/kg
      SENSITIVITY FACTOR - 2.00

RELEASE LOCATION -ABROSIMOV BAY
      RELEASE AMOUNT -   1.00 TBq
      PM CONCENTRATION - 25.00 mg/liter
      PARTITION COEFFICIENT -   .80

  RELEASE LOCATION - NOVAYA ZEMLYA TROUGH
      RELEASE AMOUNT -   1.00 TBq
      PM CONCENTRATION - 5.00 mg/liter
      PARTITION COEFFICIENT -   .95

   TIME   ABB   NZT   WKS   EKS
    .00, .21053E+04, .62524E-01, .30000E-02, .30000E-02
  365.00, .19346E+04, .62226E-01, .88001E-02, .30000E-02
1825.00, .92395E-02, .32082E-01, .88446E-02, .95744E-02
3650.00, .55041E-02, .13611E-01, .53983E-02, .75141E-02
5475.00, .40536E-02, .70027E-02, .40102E-02, .55244E-02
7300.00, .34519E-02, .45612E-02, .34336E-02, .42931E-02
9125.00, .31967E-02, .36281E-02, .31889E-02, .36324E-02
10950.00, .30866E-02, .32597E-02, .30832E-02, .33009E-02
12775.00, .30384E-02, .31098E-02, .30369E-02, .31408E-02
14600.00, .30171E-02, .30473E-02, .30165E-02, .30652E-02
16425.00, .30077E-02, .30206E-02, .30074E-02, .30299E-02
18250.00, .30035E-02, .30091E-02, .30033E-02, .30137E-02

Fig II.9. Example output file COMMA.OUT. The file contains summary information for the previous program execution that can be imported into a spreadsheet or graphics program for further analyses.

DATE = 01/18/1996
TIME = 09:44

OUTPUT FOR KARA SEA BOX MODEL

INPUT PARAMETERS

SIMULATION RUN TIME -   18250 DAYS
INSTANTANEOUS RELEASE
HALF-LIFE: .2411E+05 YRS
DISTRIBUTION COEFFICIENT (Kd) - .10E+02 m^3/kg
SENSITIVITY FACTOR - 2.00

RELEASE LOCATION -ABROSIMOV BAY
        RELEASE AMOUNT -   1.00 TBq
        PM CONCENTRATION - 25.00 mg/liter
        PARTITION COEFFICIENT -   .80

RELEASE LOCATION - NOVAYA ZEMLYA TROUGH
        RELEASE AMOUNT -   1.00 TBq
        PM CONCENTRATION -   5.00 mg/liter
        PARTITION COEFFICIENT -   .95

TIME	ABB	NZT	WKS	EKS
.00	.21053E+04	.62524E-01	.30000E-02	.30000E-02
365.00	.19346E+04	.62226E-01	.88001E-02	.30000E-02
1825.00	.92395E-02	.32082E-01	.88446E-02	.95744E-02
3650.00	.55041E-02	.13611E-01	.53983E-02	.75141E-02
5475.00	.40536E-02	.70027E-02	.40102E-02	.55244E-02
7300.00	.34519E-02	.45612E-02	.34336E-02	.42931E-02
9125.00	.31967E-02	.36281E-02	.31889E-02	.36324E-02
10950.00	.30866E-02	.32597E-02	.30832E-02	.33009E-02
12775.00	.30384E-02	.31098E-02	.30369E-02	.31408E-02
14600.00	.30171E-02	.30473E-02	.30165E-02	.30652E-02
16425.00	.30077E-02	.30206E-02	.30074E-02	.30299E-02
18250.00	.30035E-02	.30091E-02	.30033E-02	.30137E-02

Fig II.10. Example output file SPACE.OUT. The file contains summary information for the previous program execution that can be imported into a spreadsheet or graphics program for further analyses.

II.4 Program execution

II.4.1 Setup

The user should create a subdirectory on his/her computer hard drive using DOS or the Windows File Manager. Copy all of the program files on the program disk into the same subdirectory. The program is now ready for execution.

II.4.2 User prompts

The program must be executed in DOS. Exit from Windows into DOS and move to the subdirectory containing the program files. Alternatively, double-click on the DOS prompt which is normally located in the WINDOWS MAIN directory. At the subdirectory prompt type: 'KASE'. The program is running properly if a program header appears on the screen followed by, 'Specify User Level: Beginner (1) / Advanced (2)' , and the user prompt, 'ENTER User Level 1/2 -->'. The user will then be prompted for information on the type of simulation desired (see Figs. A1 and A2). Whenever a response is required from the user, the word 'ENTER' will appear as the first word of the screen prompt. The program must receive a valid response to each prompt in order to continue execution. A suggested range of values for a variable is given in parentheses at the end of the user prompt statement. After entering a response, the user must press the ENTER key. As stated previously, all numeric responses must be given as real numbers.

User level
Valid responses are '1' = Beginner or '2' = Advanced. Advanced users must provide more input options than beginners with an associated increase in control over how a model simulation is conducted.

Input mode (User Level = 2 only)
Appropriate responses are 'S' = screen or 'B' = batch. If 'S' then all input information will be requested from the user by screen prompts. If 'B' then all subsequent input information will be retrieved from the BATCH.DAT file.

Radionuclide
An appropriate response is any number 1-8 where '1' = ^{238}Pu, '2' = ^{239}Pu, '3' = ^{240}Pu, '4' = ^{134}Cs, '5' = ^{137}Cs, '6' = ^{241}Am, '7' = ^{90}Sr, '8' = ^{60}Co. Only one radionuclide may be modelled during a simulation.

the spreadsheet/graphics program to learn how to import data from outside the application.

The results of the simulations graphically displayed in Fig. II.11 demonstrate the effect of the uncertainty in K_d on ^{239}Pu activities. The effect is most pronounced at maximum ^{239}Pu activities. In all four regions, ^{239}Pu activities decrease to background (0.003 Bq/m^3) in less than 50 years. Within five years of the releases, ^{239}Pu activities peak in both the western and eastern regions of the Kara Sea.

With the exception of Abrosimov Bay, peak activities associated with the release of 1TBq of ^{239}Pu from Abrosimov Bay and the Novaya Zemlya Trough are low (< 0.1 Bq/m^3). Clearly, the choice of K_d values is critical to any further impact assessment of the movement of contaminants through the environment. If the value of K_d is assumed to be close to the highest of the recommended values, K_d = 1000 m^3/kg (IAEA 1985), the model predicts that 96% of the ^{239}Pu activity will sorb onto PM; only 6% will be transported from Abrosimov Bay into other regions of the Kara Sea in dissolved form. If the K_d value is assumed to be near the mean of the IAEA recommended values (K_d = 100 m^3/kg), then 29% of the activity will be transported in the dissolved phase. Finally, if the value of K_d is assumed to be close to the lowest recommended value (K_d = 10 m^3/kg), then 80% of the activity will remain dissolved and available for transport. Predicted ^{239}Pu activities in Abrosimov Bay at the time of discharge are 100,750 and 2100 Bq/m^3 for K_d = 1000, 100, 10 m^3/kg respectively. The results of this simulation provide additional support to the impact assessment presented in 8.7; only local scale impacts of the dumped nuclear waste pose the greatest concern to human health and the environment. Semi-enclosed, shallow areas with restricted access to the open Kara Sea tend to trap and concentrate the activity within a small volume of water, thereby posing a greater health hazard. For the nuclear waste residing in the Kara Sea, the adage, "dilution is the solution to pollution" is true.

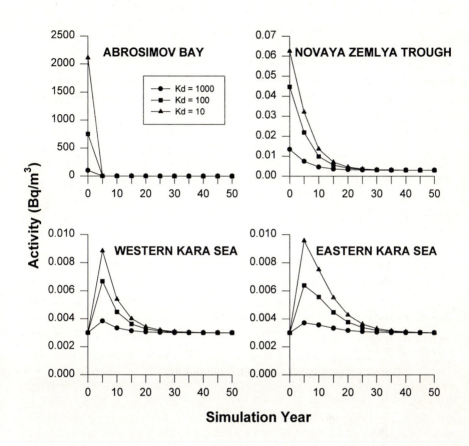

Fig II.11. Output from example simulations using the model, KASE. Simulations were conducted for K_d = 1000,100,10 m³/kg and instantaneous discharges of 1TBq ^{239}Pu from both Abrosimov Bay and the Novaya Zemlya Trough.

Appendix III. COSMO-BIO demo model

Introduction

COSMO-BIO shows the impacts of alternative user-specified coastal zone management plans for a synthetic case study are, called Whale Bay, using a set of criteria including socio-economic ones and so-called 'population risk values' and biodiversity for a number of indicator species.

Cause-effects of socio-economic activities in five different geographical sub-areas of Whale Bay are modelled on the environmental factors of water and ecosystem quality. The parameters investigated are (i) employment, gross regional product (GRP), from fishery, harbour activities, sand extraction and tourism, (ii) costs of environmental measures and (iii) a number of water quality and biodiversity criteria.

PC requirements

The DEMO of COSMO-BIO requires: PC running Windows, SVGA/VGA colour screen and 3 MB of harddisk space.

Installation

Installation: Either through Windows or Ms-Dos: A:<Setup>.

How to use

Each screen of COSMO-BIO contains an information button, which provides the user with information on how to use the programme. It is advisable to activate these buttons and read the information given in the inset on the available options and actions, before continuing to the next screen. Additional information is obtained by clicking the icons. All information is linked to a windows-help-file.

References

Aarkrog A (1994) Radioactivity in polar regions-main sources. J Environ Radioactivity 25:21-35.

Abril JM (1996) Kd distribution coefficients for radionuclides: some physical and chemical features of their variability. J Environm Radioact (in press).

Anonymous (1980) PCBs in Nederland. Centraal bureau voor de statistiek, afdeling natuurlijk milieu. Staatsuitgeverij The Hague.

Artaxo P, Vanderlei Martins J, Gerab F, Yamasoe MA (1993) Monitoring global atmospheric changes and environmental quality in tropical countries, In: Abrao JJ, Wasserman JC, Silva Filho EV, (eds) Perspectives for environmental geochemistry in tropical countries, Proc Int Symposium Niterói, Brazil, 265-269.

Asplund L (1994) Development and application of methods for determination of polychlorinated organic pollutants in biota. Thesis Stockholm University, Akademitryck, Edsbruk.

Asplund L, Svensson B-G, Nilsson A, Eriksson U, Jansson B, Jensen S, Wideqvist U, Skerfving S (1994) Polychlorinated biphenyls, 1,1,1-trichloro-2,2-bis(p-chlorophenyl)ethane (p,p'-DDT) and 1,1-dichloro-2,2-bis(p-chlorophenyl)-ethylene (p,p'-DDE) in human plasma related to fish consumption. Arch Environ Health 49:447-486.

Aston SR, Duursma EK (1973) Concentration effects on Cs-137, Zn-65, Co-60 and Ru-106 sorption by marine sediments with geochemical implications. Neth J Sea Res 6:225-240.

Badger MR (1985) Photosynthetic oxygen exchange. Ann rev Plant Physiol 36:27-58.

Baker JE, Eisenreich SJ (1990) Concentrations and fluxes of polycyclic aromatic hydrocarbons and polychlorinated biphenyls across the air-water interface in Lake Superior. Environm Scie Technol 24: 342-352.

Bakker C, Beeftink WG, Doornbos G, Duursma EK, Lambeck RHD, Lindeboom HJ, Nienhuis PH, Visscher PRM de (1984) Milieu-expertise Krammersluiting, Zoommmeer. CEMO, Yerseke, Δ-1984-4: pp 30.

Balistrieri LS, Murray JW (1981) The surface chemistry of goethite (αFeOOH) in major ion seawater. Amer J Sci 281:788-806.

Balistrieri LS, Murray JW (1983) Metal-solid interactions in the marine environment: estimating apparent equilibrium binding constants. Geochim Cosmochim Acta 47:1091-1098.

Balistrieri LS, Murray JW (1984) Marine scavenging: trace metal adsorption by interfacial sediment from MANOP Site H. Geochim Cosmochim Acta 48:921-929.

Balistrieri LS, Murray JW (1986) The surface chemistry of sediments from the Panama Basin: the influence of Mn oxides on metal adsorption. Geochim Cosmochim Acta 50:2235-2243.

Barnola JM, Raynaud D, Korotkevich YS, Lorius C (1987) Vostok ice core provides 160,000-year record of atmospheric CO_2. Nature 329:408-414.

Barua DK (1990) Suspended sediment movement in the estuary of the Ganges-Brahmaputra-Meghna river system. Mar Geol 91:243-253.

Beardall J, Mukerji D, Glover HE, Morris I (1976). The path of carbon in photosynthesis by marine phytoplankton. J Phycol 12:409-417.

Benjamin MM, Leckie JO (1980) Adsorption of metals at oxide interfaces: effects of the concentrations of adsorbate and competing metals. In: Baker RA (ed). Contaminants and sediments, vol 2:305-322.

Benjamin MM, Leckie JO (1981) Multiple-site adsorption of Cd, Cu, Zn and Pb on amorphous iron oxyhydroxide. J Colloid Interface Sci 79:209-211.

Ayotte P, Carrier C, Dewailly (1996) Health risk assessment for Inuit newborns exposed to dioxine-like compounds through breast feeding. Chemosphere: 32, 531.

Berger WH, Smetacek VS, Wefer G (1989) Ocean Productivity and Palaeoproductivity - An overview, In: Berger WH, Smetacek VS, Wefer G (eds) Productivity of the Ocean; Present and Past, John Wiley & Sons, New York, 1-34.

Berkaloff A, Bourguet J, Favard P, Favard N, Lacroix JC (1981) Biologie et Physiologie Cellulaires, III Chloroplasts, Peroxysomes, Division Cellulaire. Edition Hermann, Paris. pp 182.

Berner RA (1980) Early diagenesis. Princeton Uni Press, Princeton NJ, pp 214.

Berner RA (1991) A model for atmospheric CO_2 over phanerozoic time. Am J Sci 291:339-376.

Berner RA (1992) Weathering, plants, and the long-term carbon cycle. Geochim Cosmochim Acta 56:3225-3231.

Berner RA, Canfield DE (1989) A new model for atmospheric oxygen over phanerozoic time. Am J Sci 289:333-361.

Berner RA, Lasage AC, Garrels RM (1983) The carbonate-silicate geochemical cycle and its effect on atmospheric carbon dioxide over the past 100 million years. Am J Sci 283:641-683.

Bernard PC, Grieken RE van (1989) Geochemistry of suspended matter from the Baltic Sea. 1. Results of individual particle characterization by automated electron microprobe. Mar Chem 26:155-177.

Bernhard M (1988) Mercury in the Mediterranean. UNEP Region Seas Rep Stud No. 98, UNEP, Nairobi.

Béthoux J-P (1980) Mean water fluxes across sections in the Mediterranean Sea, evaluated on the basis of water and salt budgets and of observed salinities. Oceanol Acta 3:79-88.

Bijlsma L (ed) (1994) Preparing to meet the coastal challenges of the 21st century: Conf Rep World Coast Conference 1993. Min Transp Publ Works Water Manag, The Hague.

Black CC, Chen TM, Browns RH (1969) Biochemical basis for plant competition. J Weed Sci 17:338-348.

BNF (1981-1984) Annual reports on radioactive discharges and monitoring of the environment 1980 to 1983. British Nuclear Fuels plc. Health Safety Dir Risley. pp 87.

Brink BJE ten, Hosper SH, Colijn F (1991) A quantitative method for description and assessment of ecosystems: the Amoeba-approach. Mar Poll Bull 23:265-270.

Boer PL de (1986) Changes in the organic carbon burial during the Early Cretaceous, In: Summerhayes CP, Shackleton NJ (eds) North Atlantic Palaeoceanography, Geol Soc Spec Publ No 21:321-331.

Boëtius I, Boëtius J (1967) Studies on the European Eel, *Anguilla anguilla* (L.). Experimental induction of the male sexual cycle, its relation to temperature and other factors. Medd Dan Fisk Havunders 4:339-405.

Boisson F, Hutchins DA, Fowler SW, Fisher NS, Teyssie J-L (1996) Influence of temperature on the accumulation and retention of eleven radionuclides by the marine alga *Fucus vesiculosus* (L.). Mar Poll Bull (in press).

Bolin B, Degens ET, Duvigneaud P, Kempe S (1979) The global biogeochemical carbon cycle, In: Bolin B, Degens ET, Kempe S, Ketner P (eds) The Global Carbon Cycle, Scope 13, John Wiley & Sons, Chichester (UK), 1-56.

Bollinger MS, Moore WS (1993) Evaluation of salt marsh hydrology using radium as a tracer. Geochim Cosmochim Acta 57:2203-2212.

Boon JP, Arnhem E van, Jansen S, Kannan N, Petrick G, Schulz D, Duinker JC, Reijnders PJH, Goksøyr A (1992) The toxicokinetics of PCBs in marine mammals with special reference to possible interactions of individual congeners with the cytochrome P450-dependent monooxygenase system - An overview. Chapter 6 In: Walker CH, Livingstone DR (eds) Persistent pollutants in marine ecosystems., Pergamon Press, Oxford, 119-159.

Boon JP, Eijgenraam F, Everaarts JM, Duinker JC (1989) A structure-activity relationship (SAR) approach towards metabolism of PCBs in marine animals from different trophic levels. Mar Environ Res 27:159-176.

Boon JP, Oostingh I, Meer J van der, Hillebrand TJ (1994) A model for the bioaccumulation of chlorobiphenyl congeners in marine mammals. Eur J Pharmacol, Environ Toxicol Pharmacol Sect, 270:237-251.

Bouwman AF, Born GJ van de, Swart RJ (1992) Land-use related sources of CO_2, CH_4 and N_2O; current global emissions and projections for the period 1990-2100. Report 222901004, Nat Inst Public Health & Environment Protection, Bilthoven (Nl), pp 102.

Bowden KF (1965) Currents and mixing in the ocean, In: Riley JP, Skirrow G (ed) Chemical Oceanography, 1rst ed., Academic Press, London, vol. I:43-72.

Boyle EA (1976) The marine geochemistry of trace metals. PH.D. thesis. Woods Hole Oceanographic Inst./MIT.

Broecker WS (1970) Man's oxygen reserves. Science 168:1537-1538.

Broecker WS (1990) Comment on "iron deficiency limits phytoplankton growth in Antarctic waters" by John H. Martin et al. Glob Biochem Cycl 4:3-4.

Bromley DW, Cernea MM (1989) The management of common property natural resources. World Bank discussion papers 57: pp 66.

Brownawell BJ, Farrington (1985) Partitioning of PCBs in marine sediments. In: Marine and estuarine geochemistry, Sigleo AC, Hattori A (eds) Lewis Publ Chapt 7:1-24.

Brunn H (1992) Exposition Nahrung Teil II: Fremdstoffe in Frauenmilch - chlorierte Kohlenwasserstoffe. Ernährungs Umschau 39:369-374.

Bruggeman WA (1982) Hydrophobic interactions in the aquatic environment. In: The Handbook of Environmental Chemistry, Hutzinger O (ed), Springer, Berlin Heidelberg New York, Vol. 2/Part B, 29-48.

Buch K, Harvey HW, Wattenberg H, Grippenberg S (1932) Ueber das Kohlensäuresystem in Meerwasser Rapp et Proces-Verb des Réunions du Conseil Perm Int p l'Explor de la Mer Vol LXXIX:1-70.

Budyko MI, Ronov AB, Yanshin AL (1987) History of the Earth's Atmosphere, Springer, Heidelberg, Berlin, New York, pp 139.

Buesseler KO, Sholkovitz ER (1987a) The geochemistry of fallout plutonium in the North Atlantic: I. a pore water study in shelf, slope and deep-sea sediments. Geochim Cosmochim Acta 51:2605-2622.

Buesseler KO, Sholkovitz ER (1987b) The geochemistry of fallout plutonium in the North Atlantic: II. $^{240}Pu/^{239}Pu$ ratios and their significance. Geochim Cosmochim Acta 51:2623-2637.

Burkhard LP, Andren AW, Armstrong DE (1985a) Estimation of vapor pressures for polychlorinated biphenyls: a comparison of eleven predictive methods. Environ Sci Technol 19:500-506.

Burkhard LP, Armstrong DE, Andren AW (1985b) Partitioning behavior of polychlorinated biphenyls. Chosphere 14:1703-1715.

Burkhard LP, Armstrong DE, Andren AW (1985c) Henry's law constants for polychlorinated biphenyls. Environ Sci Technol 19:590-506-596.

Caldeira K, Kasting JF (1992) The life span of the biosphere revisited. Nature 360:731-723.

Calmet D, Guegueniat P (1985) Les rejets d'effluents liquides radioactifs du centre de traitement des combustibles irradiés de la Hague (France) et l'évolution radiologique du domaine marin. IAEA, Vienna. TECDOC-329:111-144.

Carroll J, Boisson F, Fowler SW, Teyssie J-L (1996) Radionuclide adsorption to sediments from nuclear waste dumping sites in the Kara Sea. Mar Poll Bull (in press).

Carroll J, Falkner KK, Brown ET, Moore WS (1993) The role of the Ganges-Brahmaputra mixing zone in supplying barium and ^{226}Ra to the Bay of Bengal. Geochim Cosmochim Acta 57:2981-2990.

Carroll J, Lerche I (1996) A note on partition coefficient distributions. J Math Geol (in press).

Carroll J, Lerche I, Abraham JD, Cisar DJ (1995) Model-determined sediment ages from ^{210}Pb profiles in un-mixed sediments. J Geophys Res 9:553-565.

Carson R (1962) Silent spring. Houghton Mifflin Comp Boston, The Riverside Press, Cambridge, pp. 368.

Chen S, Eisma D, Kalf J (1994) *In situ* size distribution of suspended matter during the tidal cycle in the Elbe estuary. Neth J Sea Res 32: 37-48.

Chou L, Wollast R (1996) Biogechemical behavior and mass balance of dissolved aluminum in the Western Mediterranean Sea. Special Issue: Functioning of the Western Mediterranean Sea: present and future. Deep Sea Res (in press).

Coleman JM (1969) Brahmaputra River: Channel processes and sedimentation. Sediment Geol 3:129-239.

Collins RP, Jones MB (1985) The influence of climatic factors on the distribution of C_4 species in Europe. Vegetatio 64:121-129.

Copin-Montégut C (1988) Eléments majeurs des particules en suspension de la Méditerranée occidentale. Oceanol Acta Spec Issue 9:95-102.

Côté B, Platt T (1983) Day-to-day variations in the spring-summer photosynthetic parameters of coastal marine phytoplankton. Limnol Oceanogr 28:320-344.

Cox RA (1965) The physical properties of seawater In: Chemical Oceanography, Riley JP, Skirrow G (eds.) Acad Press 1rst ed. p 73-120.

Cruise Report (1994) Radioactive contamination at dumping sites for nuclear waste in the Kara Sea. Joint Russian-Norwegian Expert Group for Investigation of Radioactive Contamination in the Northern Areas. Norwegian Radiation Protection Authority, Norway.

Cullen JJ (1995) Status of the iron hypothesis after the open ocean enrichment experiment. Limnol Oceanogr 40:1336-1343.

Dai MH, Martin JM (1995) First data on trace metal level and behaviour in two major Arctic river-estuarine systems (Ob and Yenisey) and in the adjacent Kara Sea, Russia. Earth Planet Sci Lett 131:127-141.

Davis JA (1984) Complexation of trace metals by adsorbed natural organic matter. Geochim Cosmochim Acta 48:679-692.

Dawson R, Duursma EK (1974) Distribution of radioisotopes between phytoplankton, sediment and sea water in a dialysis compartment system. Neth J Sea Res 8: 339-353.

Degani G, Gallagher ML, Meltzer A (1989) The influence of body size and temperature on oxygen consumption of the European eel, *Anguilla anguilla*. J Fish Biol 34:19-24.

Degens ET (1982) Geochemical balance of organic matter and Geochemical cycle, In: McGraw-Hill Encyclopedia of Science and Technology, McGraw-Hill, London, Vol 9:636-638.

Degens ET (1989) Perspectives on biogeochemistry. Springer, Berlin Heidelberg New York, pp 423.

De Luca Rebello A, Moreira I (1982) The influence of various seawater components on the buffer capacity for CO_2. Mar Chem 11:33-41.

Descolas-Gros C, Fontugne M (1990) Stable carbon isotope fractionation by marine phytoplankton during photosynthesis. Plant Cell Environ 13:207-218.

Dietrich G, Kalle, K, Krause, W, Siedler G (1975) Allgemeine Meereskunde. Gebr. Borntraeger, Berlin, Stutgart, 3 ed. pp 593.

Dietrich G, Kalle, K, Krause, W, Siedler G (1980) General oceanography. Wiley Intersc Publ New York, Chichester, Brisbane, Toronto, 2ed. pp 626.

DiToro DM, Paquin PR (1984) Time variable model of the fate of DDE and Lindane in a quarry. J Soc Environ Toxicol Chem 3:335-353.

Doskey PV, Andren AW (1981) Modelling the flux of atmospheric polychlorinated biphenyls across the air/water interface. Am Chem Soc 15:705-711.

Dugdale RC, Wilkerson FP (1990) Iron addition experiments in the Antarctic: a reanalysis. Glob Biochem Cycl 4:13-19.

Duinker JC, Bouchertall F (1989) On the distribution of atmospheric polychlorinated biphenyl congeners between vapor phase, aerosols and rain. Environ Sci Technol 23:57-62.

Duinker JC, Schulz DE, Petrick G (1988) Multidimensional gaschromatography with electron capture detection for the determination of toxic congeners in polychlorinated biphenyl mixtures. Anal Chem 60:478-482.

Duursma EK (1961) Dissolved organic carbon, nitrogen and phosphorus in the sea. Neth J Sea Res 1:1-148.

Duursma EK (1965) The dissolved organic constituents of sea water. Chapter 11 in: Riley JP, Skirrow G (eds) Chemical Oceanography, Vol. I (1rst edition) Acad Press, London 433-475.

Duursma EK (1970) Organic chelation of Co-60 and Zn-65 by leucine in relation to sorption by sediments. In: Organic Matter in Natural Waters, Symp. Alaska, Hood DW, (ed.) Publ. Inst Mar Sci Univ Alaska, no. 1:387-397.

Duursma EK (1972) Geochemical aspects and applications of radionuclides in the sea. (An extensive literature review). Oceanogr Mar Biol Ann Rev 10:137-223.

Duursma EK (1973) Specific activity of radionuclides sorbed by marine sediments in relation to the stable element composition. In: Radioactive Contamination of the Marine Environment, IAEA, Vienna, SM. 158/4:57-71.

Duursma EK (1976) Role of pollution and pesticides in brackish water aquaculture in Indonesia. FAO, FI:INS/72/003/4, W/H9164:1-42.

Duursma EK (1977) Effect of hydrostatic pressure on the diffusion of chloride in marine sediment. An *in situ* experiment. Deep-Sea Res 24:1161-1166.

Duursma EK, Arthurton RS, Latouche C (eds) (1996) Coastal change 95. BORDOMER-IOC, Bordeaux-Paris. Int Conf Bordeaux 1995, pp 1023.

Duursma EK, Bewers JM (1986) Application of Kd's in marine geochemistry and environmental assessment. In: Application of Distribution Coefficients to Radiological Assessment Models. Sibley THM, Myttenaere C (eds). Elsevier Appl Sci Publ London, pp 138-165.

Duursma EK, Bosch CJ (1970) Theoretical, experimental and field studies of radioisotopes concerning diffusion in sediments and suspended particles in the sea. Part B. Methods and Experiments. Neth J Sea Res 4:395-469.

Duursma EK, Boisson MPRM (1994) Global oceanic and atmospheric oxygen stability considered in relation to the carbon cycle and to different time scales. Oceanol Acta 17:117-141.

Dewailly E (1996) Polychlorinated biphenyl (PCB) and dichlorodiphenyl-diethylene (DDE) concentrations in breast milk of woman in Quebec. Am J Publ Health: 86, 1241.

Duursma EK, Dawson R (1975) Competition and time of sorption for various radionuclides and trace metals by marine sediments and diatoms. Thal Jugosl 11:47-51.

Duursma EK, Dawson R (eds) (1981) Marine Organic Chemistry. Elsevier Sc Publ Co, Amsterdam, pp 521.

Duursma EK, Dawson R, Ros Vicent J (1975) Competition and time of sorption for various radionuclides and trace metals by marine sediments and diatoms. Thalassia Jugosl 11:47-52.

Duursma EK, Eisma D (1973) Theoretical, experimental and field studies concerning reactions of radioisotopes with sediments and suspended particles of the sea. Part C: Applications to Field Studies. Neth J Sea Res 6: 265-324.

Duursma EK, Engel H, Martens TJM (eds) (1982) De Nederlandse Delta; een compromis tussen milieu en techniek in de strijd tegen het water. Natuur en Techniek, Maastricht, pp. 511. (with English abstract of 48 pp and map; 550 colour illustrations).

Duursma EK, Frissel MJ, Guary JC, Martin JM, Nieuwenhuize J, Pennders RMJ, Thomas AJ (1985) Plutonium in sediments and mussels of the Rhine-Meuse-Scheldt estuary. In: Proc. CEC Seminar on the behaviour of radionuclides in estuaries, Renesse, Sept. 1984, Luxembourg, 71-106.

Duursma EK, Geldermalsen LA van, Wegereef JW (1983) Migration processes in marine sediments caused by heat sources: simulation experiments related to deep sea disposal of high level radioactive wastes. (Eur. 8710), European Appl Res Rep Nucl Sci Technol 5:(3) 451-512.

Duursma EK, Hoede C (1967) Theoretical, experimental and field studies concerning molecular diffusion of radioisotopes in sediments and suspended solid particles of the sea. Part A: Theories and mathematical calculations. Neth J Sea Res 3: 423-457.

Duursma EK, Marchand M (1974) Aspects of organic marine pollution. (An extensive literature review). Oceanogr Mar Biol Ann Rev 12:315-431.

Duursma EK, Nieuwenhuize J, Liere JM van (1983) Organochlorine contamination of the Dutch Delta Region as "watched" by mussels. Wat Sci Techn 16:619-622.

Duursma EK, Nieuwenhuize J, Liere JM van (1989) PCB equilibria in an estuarine system. Scienc Tot Environm 79:141-155.

Duursma EK, Nieuwenhuize J, Liere JM van, Hillebrand T (1986) Partitioning of organochlorines between water, particulate matter and some organisms of the Rhine-Meuse--Scheldt Delta. Neth J Sea Res 20:239-251.

Duursma EK, Nieuwenhuize J, Liere JM van, Rooy CM de, Witte JIJ, Meer J van der (1991) Possible loss of PCBs from migrating European silver eels; a 3-month simulation experiment. Mar Chem 36:215-232.

Duursma EK, Parsi P (1976) Distribution of plutonium-237 between sediment and sea water. Rapp Comm int Mer Médit 23:159-160.

Duursma EK, Ruardij P (1989) Conservative mixing in estuaries as affected by sorption, complexing and turbidity maximum: a simple model example. Mar Chem 28:251-258.

Dyal RS, Hendricks SB (1950) Total surface of clays in polar liquids as a characteristic index. Soil Sci 62: 421-432.

Edmond JM, Boyle ED, Drummond D, Grant B, and Gislik T (1978) Desorption of barium in the plume of the Zaire (Congo) River. Neth J Sea Res 12:324-328.

Edwards GE, Walker D (1983) C_3, C_4: mechanisms, and cellular and environmental regulation, of photosynthesis. Blackwell Scientific Publications, Oxford (UK), pp 409.

Ehleringer JR, (1978) Implications of quantum yield differences on the distributions of C_3 and C_4 grasses. Oecologia 31:255-267.

Eisenreich SJ, Johnson TC (1983) PCBs in the great lakes: sources, sinks, burdens. In: D'Itri FM, Kamrin MA (eds). PCBs: human and environmental hazards 4:49-75.

Eisma D (1991) Particle size of suspended matter in estuaries. Geo-Mar Letters 11: 147-153.

Elsinger RJ, Moore WS (1980) Ra-226 behaviour in the Pee Dee-Winyah Bay Estuary. Earth Planet Sci Lett 48:239-249.

Erickson MD (1986) Analytical chemistry of PCBs. Butterworth Publ. Boston, London pp 457.

Erickson RJ, McKim JM (1990) A simple flow-limited model for exchange of organic chemicals at fish gills. Environ Toxicol Chem 9:159-165.

ESA (1992) Atmosphere quality standards in manned space vehicles. European Space Agency, ESTEC, Noordwijk, Netherlands, ESA PSS03-401, Issue 1.

Etcheto J, Merlivat L (1988) Satellite determination of the carbon dioxide exchange coefficient at the ocean-atmosphere interface: a first step. J Geoph Res 93:669-678.

Eysink WD (1983) Basic consideration on the morphology and land accretion potentials in the estuary of the lower Meghna River; Land Reclamation Project, Bangladesh Water Development Board, Rept. 15.

Fonds M, Jaworski A, Iedema A, Puyl PVD (1989) Metabolism, food consumption, growth and food conversion of Shorthorn Sculpin (*Myoxocephalus scorptius*) and Eelpout (*Zoarces vivparus*). ICES, Demersal Fish Comm CM 1989/G:31 pp 19.

Foth HD, Ellis BG (1988) Soil fertility. J Wiley, New York, pp 212.

Fowler SW, Baxter MS, Hamilton TF, Miquel JC, Osvath I, Povinec PP, Scott EM (1994) International radiological assessment programs related to radioactive waste dumping in the Arctic Sea: IAEA-MEL's role. Proceedings Workshop Arctic Contamination, May 2-7 1993. Anchorage AK, Int Arct Res Policy Comm.

Fyfe WS (1992) Geosphere interactions on a convecting planet: mixing and separation, In: Hutzinger O (ed) The Natural Environment and the Biogeochemical Cycles, The Handbook of Environmental Chemistry, vol. 1, part F, Springer, Heidelberg, Berlin, New York, 1-26.

Garrels RM, Perry EA (1974) Cycling of carbon, sulfur and oxygen through geologic time, In: Goldberg E (ed) The Sea, J Wiley, New York, vol 5:303-316.

Geldermalsen LA van, Wegereef JW (1985) The influence of temperature on migration of radionuclides in deep-sea sediments. Simulation experiments concerning sorption and heat flow related to deep-sea disposal of high-level radioactive waste. TU Twente, TW 85/TA-Z/025 JFH, CEMO Yerseke Netherlands, Comm. Δ-305 pp 137.

Glasstone S (1960) Textbook of physical chemistry. McMillan and Co., Ltd, London, 2nd ed. pp 1320.

Goldstein JA, Safe S (1989) Mechanism of action and structure-activity relationships for the chlorinated dibenzo-*p*-dioxins and related compounds. In: Kimbrayh RD, Jensen AA (eds) Halogenated biphenyls, triphenyls, naphtalenes, dibenzodioxins and related compounds. Topics in environmental health 4, Chap 9, 2nd ed Elsevier Amsterdam, pp 239-293.

Goudriaan J, Ajtay GL (1979) The possible effects of increased CO_2 on photosynthesis, In: Bolin B, Degens ET, Kempe S, Ketner P (eds) The Global Carbon Cycle, Scope 13, John Wiley & Sons, Chichester (UK), 237-249.

Grim RE (1953) Clay mineralogy. McGraw-Hill, London, pp 384.

Groen P (1974) De Wateren van de Wereldzee. De Boer Maritieme Handboeken, Bussum, (Nl), pp 363.

Guegueniat P, Le Hir P (1981) Données nouvelles sur la dispersion des radionucléides dans la Manche. In: Impacts of radionuclide release into the marine environment. IAEA Vienna SM 248/120:481-499.

Hall DO, Rao KK (1987) Photosynthesis. New Studies in Biology, Edward Arnold, 4th ed., pp 122.

Hamilton TF, Ballestra S, Baxter MS, Gastaud J, Osvath I, Parsi P, Povinec PP (1994) Radiometric investigations of Kara Sea sediments and preliminary radiological assessment related to dumping of radioactive wastes in the Arctic Seas. J Environ Radioact 25:113-134.

Hardy EP, Krey PW, Volchok HL (1973) Global inventory and distribution of fallout plutonium. Nature 241:444-445.

Harding GC, Leblanc RJ, Vass WP, Addison, RF, Hargrave BT, Pearre S, Dupuis A, Brodie PF (in press) Bioaccumulation of polychlorinated biphenyls (PCBs) in the marine pelagic food web, based on a seasonal study in the southern Gulf of St. Lawrence, 1967-77. Mar Chem (in press).

Harms IH (1992) A numerical study of the barotropic circulation in the Barents and Kara Seas. Cont Shelf Res 12:1043-1058.

Harms IH (1995) Local and regional scale dispersion scenarios of ^{137}Cs and ^{239}Pu released from waste dumped in the Kara Sea. In: Environmental radioactivity in the Arctic. Proc Int Conf Oslo Norway 21-25 Aug 1995, Norwegian Radiation Protection Authority, Osteras Norway.

Harms IH (1996) Modeling the dispersion of ^{137}Cs and ^{239}Pu released from dumped waste in the Kara Sea. J Mar Syst (in press)

Harms IH, Backhaus JO (1992) Numerical dispersion studies of passive tracers in the Barents and Kara Seas. In: Proc 2nd Int Offshore and Polar Engineering Conf, San Francisco 1:31-37.

Harrison WG, Platt T (1980) Variations in assimilation number of coastal marine phytoplankton: effects of environmental co-variates. J Plankton Res 2:249-260.

Hayton WL, Barron MG (1990) Rate-limiting barriers to xenobiotic uptake by the gill. Environ Toxocol Chem 9: 151-157.

Hofstra JJ, Stienstra AW (1977) Growth and photosynthesis of closely related C_3 and C_4 grasses, as influenced by light intensity and water supply. Acta Bot Neerl 26:63-72.

Holland D (1978) The Chemistry of the Atmosphere and Oceans, John Wiley & Sons, New York, pp 351.

Honeyman BD, Balistrieri LS, Murray JW (1988) Oceanic trace metal scavenging: the importance of particle concentration. Deep-Sea Res 35: 227-246.

Houghton RA, Skole DL, Lefkowitz DS (1991) Changes in the landscape of Latin America between 1850 and 1985. II Net release of CO_2 to the atmosphere. Forest Ecology and Management 38:173-199.

Hutchins DA, Teyssie J-L, Boisson F, Fowler SW, Fisher NS (1995). Temperature effects on uptake and retention of contaminant radionuclides and trace metals by the brittle star *Ophiothrix fragilis* Mar Environ Res (in press).

IAEA (1984) The oceanographic and radiological basis for the definition of high-level wastes, unsuitable for dumping at sea. IAEA Safety Series, No. 66, IAEA, Vienna. 50 pp.

IAEA (1985) Sediment K_ds and concentration factors for radionuclides in the marine environment. IAEA Tech Rep No 247, IAEA Vienna, pp 73.

IAEA (1986a) World survey of isotope concentration in precipitation (1980-1983). Environmental isotope data No. 8. Technical Report Series 264, IAEA, Vienna, pp 184.

IAEA (1986b) An oceanographic model for the dispersion of wastes disposed of in the deep sea. Technical Report Series 263, IAEA Vienna, pp 166.

IAEA (1992) Radioactive waste management. An IAEA source book. IAEA, Vienna, pp 276.

IAEA-MEL (1994) Programmes related to the radioactive waste dumped in the Arctic Seas. In: IAEA-MEL Report of Recent Activities. Vienna:22-38.

References   267

IAEA (1995) Highlights of activities. IAEA/PI/A22E 95-02258, IAEA, Vienna, pp 87.
IMW (1994) International Mussel Watch project. Final report initial implementation phase. Woods Hole, Mass., pp 63.
Ingle SE, Culverson CH, Hawley JE, Pytkowicz RM (1973) The solubility of calcite in seawater at atmospheric pressure and 35 ‰ salinity. Mar Chem 1:295-307.
Iwata H, Tanabe N, Sakai N, Tatsukawa R (1993) Distribution of persistent organochlorines in the oceanic air and surface sea water and the role of ocean on their global transport and fate. Env Sci Technol 27: 1080-1098.
Jannasch HW, Honeyman BD, Balistrieri LS, Murray JW (1988) Kinetics of trace element uptake by marine particles. Geochim Cosmochim Acta 52:567-577.
Jong APJM de, Marsman JA, Hartog RS den, Hijman WC, Boer AC den, Liem AKD (1994) WHO coordinated proficiency study for the determination of polychlorinated biphenyls, dibenzodioxins and dibenzofurans in human milk and blood, cow's milk and fish. Results of the participation of RIVM-LOC in the IIIrd round. RIVM Rep 639102.009, Bilthoven, Netherlands, pp 17.
Kabata-Pendia A, Pendia H (1984) Trace elements in soils and plants. CRC Press Inc. Boca Raton, USA pp 315.
Kanja LW, Skaare JU, Ojwang SBO, Marai CK (1992) A comparison of organochlorine pesticide residues in maternal adipose tissue, maternal blood, cord blood and human milk from mother/infant pairs. Arch Environ Contam Toxicol 22:21-24.
Keeling RF, Bender ML, Tans PP (1993) What atmospheric oxygen measurements can tell us about the global carbon cycle. Global Biochem Cycles 7:37-67.
Keeling RF, Shertz SR (1992) Seasonal and interannual variations in atmospheric oxygen and implications for the global carbon cycle. Nature 358:723-727.
Kennish MJ (1989) Practical handbook of marine science. CRC Press, Boca Raton, USA, pp 710.
Key RM, Stallard RF, Moore WS, Sarmiento JL (1985) Distribution and flux of ^{226}Ra and ^{228}Ra in the Amazon River estuary. J Geophys Res 90:6995-7004.
Klasinc L, Cvitas T (1996) The photosmog problem in the Mediterranean region. Mar Chem (submitted).
Kump LR (1992) Oxygen, biochemical cycle. Encyclop Earth Sci 3:515-524.
Kump LR (1993) The coupling of the carbon and sulfur biogeochemical cycles over Phanerozoic time, In: Wollast R, Mackenzie FT, Chou L (eds) Interaction of C, N, P and S Biochemical Cycles and Global Change, Springer, Heidelberg, Berlin, New York, 475-490.
Kump LR, Holland HD (1992) Iron in precambrian rocks: implication for the global oxygen budget of the ancient earth. Geochim Cosmochim Acta 56:3217-3233.
Larcher W (1983) Physiological Plant Ecology, Springer, Heidelberg, Berlin, New York, pp 303.
Larsson P, Jänmark C, Södergren A (1992) PCBs and chlorinated pesticides in the atmosphere and aquatic organisms of Ross Island, Antarctica. Mar Poll Bull 25:281-287.
Leo A, Hansch C, Elkins D (1971) Partition coefficients and their uses. Chem Rev 71:525-616.
Li YH (1981) Ultimate removal mechanisms of elements from the ocean. Geochim Cosmochim Acta 45:1659-1664.
Li YH, Burkhardt, L, Buchholtz M, O'Hara P, Santschi PH (1984a) Partition of radiotracers between suspended particles and seawater. Geochim Cosmochim Acta 48:2011-2019.
Li YH, Burkhardt L, Teraoka H (1984b) Desorption and coagulation of trace elements during estuarine mixing. Geochim Cosmochim Acta 48:1879-1884.

Li YH, Chan LH (1979) Desorption of Bay and ^{226}Ra from river-borne sediments in the Hudson estuary. Earth Planet Sci Lett 43:343-350.

Liu J, Carroll J, Lerche I (1991) A technique for disentangling temporal source and sediment variations from radioactive isotope measurements with depth. Nucl Geophys 4:31-45.

Lide D (Ch. ed) (1993) Handbook of chemistry and physics. CRC Press, Boca Raton, USA, 74th edition.

Luther GW (1990) The frontier-molecular-orbital theory approach in geochemical processes. In: Stumm W (ed). Aquatic chemical kinetics, Wiley Interscience, New York, 173-198.

Luther GW, Kostka JE, Church TM, Sulzberger B, Stumm W (1992) Seasonal iron cycling in the marine environment: the importance of ligand complexes with Fe(II) and Fe(III) in the dissolution of Fe(III) minerals and pyrite, respectively. Mar Chem 40:81-103

Luykx F, Fraser G (1980, 1983) Radioactive effluents from nuclear power stations and nuclear fuel reprocessing plants in the European Community, discharge data 1974-1978 and 1976-1980, CEC Reports, Luxembourg, pp 55.

Lyman J (1956) Thesis, University of California, cited by Skirrow (1965).

Mackay D (1982) Solubility, partition coefficients, volatility and evaporation rates. In: Hutzinger O (ed.) The Handbook of Environmental Chemistry, Vol. 2 Part. A, Reactions and processes, Springer Berlin Heidelberg New York, 31-45.

Martin JH, Fitzwater SE (1988) Iron deficiency limits phytoplankton growth in the north-east Pacific subarctic. Nature 331:341-343.

Martin JH, Fitzwater SE, Gordon RM (1990) Iron deficiency limits phytoplankton growth in Antarctic waters. Glob Biochem Cycl 4:5-12.

Martin JM (1992) Introduction to EROS 2000. In: Martin J-M, Barth H (eds) European river ocean system. CEC Water Poll Res Rep 28:3-5.

Martin JM, Elbaz-Poulichet F, Guieu C, Loÿe-Pilot MD, Han G (1989) River versus atmospheric input of material to the Mediterranean Sea: and overview. Mar Chem 28:159-182.

Martin JM, Thomas AJ (1990) Origins, concentrations and distributions of artificial radionuclides discharged by the Rhône River to the Mediterranean Sea. J Environ Radioact 11:105-139.

Martin JM, Wollast R, Loÿens M, Thomas AJ, Mouchel JM, Nieuwenhuize J (1994) Origin and fate of artificial radionuclides in the Scheldt estuary. Mar Chem 46:189-202.

Matuo YK, Lopes JN, Casanova IC, Matuo T, Lopes JLC (1992) Organochlorine pesticide residues in human milk in the Ribeirão Preto Region, State of São Paulo, Brazil. Arch Environ Contam Toxicol 22:167-175.

McElroy MB (1983) Marine biological controls on atmospheric CO_2 and climate. Nature 302:328-329.

Meent D van de, Hollander HA, Verboom JH (1991) Sorption kinetics of micropollutants from suspended particles: experimental observations and modelling. In: Angeletti G, Bjørseth A (eds). Organic micropollutants in the aquatic environment. Proc 6th Europ Symp Lisbon, Kluwer Dordrecht, 50-60.

Merlivat L, Etcheto J, Noutin J (1991) CO_2 exchange at the air-sea interface: time and space variability. Atm Space Res Cospar Proc 11:77-85.

Millero FJ (ed) (1993) Marine physical chemistry; in memory of the contributions made to the field by Dr. Ricardo Pytkowicz, Mar Chem 44:105-280 (13 papers).

Millero FJ, Sohn ML (1991) Chemical Oceanography. CRC Press, London, pp 531.

Milliman JD (1994) Organic matter content in U.S. Atlantic continental slope sediments: decoupling the grain-size factor. Deep-Sea Res 41: 797-808.

Milliman J, Meade RH (1983) World-wide delivery of river sediments to the oceans. J Geol 91:1-21.

Min Environm (1995) Qualité de l'air en France, Bilan 1994. Serv. de l'Environnement Industriel, Ministère de l'Environnement, Paris, juillet, 1995 pp 58.

Miquel JC (1996) Environment and biology of the Kara Sea: a general overview for contamination studies. Mar Poll Bull (In press).

Molero J, Sachez-Cabeza JA, Merino J, Vives Batlle J, Mitchell PI, Vidal-Quadras A (1995) Particulate distribution of plutonium and americium in surface waters from the Spanish Mediterranean coast. J Environ Radioactivity 28:271-283.

Morel FF, Gschwend PM (1987) The role of colloids in the partitioning of solutes in natural waters, In: Stumm W (ed) Aquatic surface chemistry: chemical processes at the particle-water interface, John Wiley, New York (NY), 405-422.

Morel F, McDuff RE, Morgan JJ (1976) Theory of interaction intensities, buffer capacities, and pH stability in aqueous systems, with application to the pH of seawater and a heterogeneous model ocean system. Mar Chem 4:1-28.

Mount ME, Sheaffer MK, Abbott DT (1993) Estimated inventory of radionuclides in former Soviet Union naval reactors dumped into the Kara Sea. In: Proc environmental radioactivity in the Arctic and Antarctic (NRPA), Osteras, Norway, 81-87.

NDRE (1995) Proceedings from the workshop on modelling requirements for water mass dynamics, ice and river transports in the Kara Sea 26-30 June, Norwegian Defence Research Establishment Tjome, Norway.

NEA-OECD (1984) Seabed disposal of high-level radioactive waste. A status report on the NEA coordinated research programme. OECD, Paris,pp 247.

Nieuwenhuize J, Liere JM van (1982) Organochloorverbindingen in enkele monsters moedermelk. Report CEMO, Yerseke, Netherlands, DIHO-BL-1982-1 pp 7.

NRDC (1995) Known nuclear tests worldwide, 1945-1994. NRDC Nuclear notebook, Bull Atomic Scientists May/June 1995, 70-71.

Nyffeler UP, Li Y-H, Santschi PH (1984) A kinetic approach to describe trace-element distribution between particles and solution in natural aquatic systems. Geochim Cosmochim Acta 48:1513-1522.

ONR (1995) Proceedings from the workshop on the arctic nuclear waste assessment program 1-4 May 1995, Office of Naval Research Woods Hole Oceanographic Institution, MA, USA.

Osvath I, Ballestra S, Baxter MS, Gastaud J, Hamilton TF, Harms IH, Liong Wee Kwong L, Parsi P, Povinec PP (1995) IAEA-MEL's contribution to the investigation of the Kara Sea dumping sites. Proc Intern Conf Envir Radioact in the Arctic. Oslo Norway 21-25 August. Norwegian Radiation Protection Authority, Ostera, Norway.

OECD-NEA (1985) Review of the continued suitability of the dumping site for radioactive waste in the North-East Atlantic. NEA-OECD, Paris.

Oppo DW, Fairbanks RG (1990) Atlantic Ocean thermohalene circulation of the last 150,000 years: relationship to climate and atmospheric CO_2. Palaeoceanography 5:277-288.

Paluszkiewicz T, Hibler LF, Richmond MC, Bradley DJ, Thomas SA (1996) Modelling the potential radionuclide transport by the Ob and Yenisey Rivers to the Kara Sea. Mar Poll Bull (in press).

Paustenbach DJ (1989) A survey of health risk assessment. In: Paustenbach DJ (ed) The risk assessment of environmental hazards. John Wiley & Sons, New York: 27-124.

Pentreath RJ (1973) The roles of food and water in the accumulation of radionuclides by marine teleost and elasmobranch fish. In: Radioactive contamination of the marine environment. IAEA-SM-158/26, 421-436.

Pentreath RJ (1978) ^{237}Pu experiments with the plaice *Pleuronectes platessa*. Mar Biol 48:327-335.

Pentreath RJ (1985) Radioactive discharges from Sellafield (UK). IAEA, Vienna. TECDOC-329:67-110.

Pessenda LCR, Valencia EPE, Telles ECC, Cerri CC, Camargo PB, Martinelli LA (1993) The association C-14 dating and δC-13 in palaeoecology studies in Brazilian tropical and subtropical soils. In: Abrão JJ, Wasserman JC, Silva Filho EV (eds) Perspectives for environmental geochemistry in tropical countries, Proc Int Symp Niterói, Brazil, 89-91.

Plath DC, Pytkowicz RM (1980) The solubility of aragonite at 25.0 °C and 32.62 ‰ salinity. Mar Chem 10:3-7.

Plath DC, Johnson KS, Pytkowicz RM (1980) The solubility of calcite - probably containing magnesium - in seawater. Mar Chem 10:9-29.

Pond S and Pickard GL (1983) Introductory dynamical oceanography. Pergamon Press, Oxford. pp 329.

Povinec PP, Osvath I, Baxter MS (1995) Marine scientists on the Arctic Seas: documenting the radiological record. IAEA Bulletin, Vienna Austria, 37:31-35.

Povinec PP, Osvath I, Baxter MS, Ballestra S, Carroll J, Gastaud J, Harms I, Huynh-Ngoc L, Liong Wee Kwong L, Pettersson H (199?) IAEA-MEL's contribution to the investigation of Kara Sea radioactivity and radiological assessment. Mar Poll Bull (in press).

Quay PD, Tilbrook B, Wong CS (1992) Oceanic uptake of fossil fuel CO_2: carbon-13 evidence. Science 256:74-79.

Rapaport RA, Eisenreich SJ (1984) Chromatographic determination of octanol-water partition coefficients (K_{ow}'s) for 58 polychlorinated biphenyl congeners. Environ Sci Technol 18:163-170.

Rijsberman FR, Westmacott RS, Waardenburg D (1995) CORONA: Coastal resources management roleplay. Trainers' manual. Res Anal, Delft and RIKZ, Nat Inst Coast Mar Manag, The Hague.

RIME (1971) Radioactivity in the Marine Environment. US Nat Acad Sc Washington, D.C. pp 272.

Ros Vicent J, Costa Yangue F, Parsi P, Statham G, Duursma EK (1974) The ease of release of some trace metals and radionuclides being sorbed for prolonged periods by marine sediments. Report to IAEA pp 12 (see also Duursma et al. 1975).

Roy RN, Roy LN, Vogel KM, Porter-Moore C, Pearson T, Good CE, Millero FJ, Campbell DM (1993) The dissociation constants of carbonic acid in seawater at salinities 5 to 45 and temperatures 0 to 45 °C. Mar Chem 44:249-267.

Santschi PH, Nyffeler UP, Li Y-H, and O'Hara P (1986). Radionuclide cycling in natural waters: relevance of scavenging kinetics. In: Sly P (ed) Interactions between sediments and water, Springer, Heidelberg, Berlin, New York:183-191.

Sazykina TG, Kryshev IL (1994) Provision of site specific input data for an assessment of radiological impact of waste dumping in to the Barents and Kara Seas. IAEA-IASAP working paper 4:1-93.

Schindler PW (1975) Removal of trace metals from the oceans: a zero order model. Thalassia Yugoslavica 11:101-111.

Schmitt MR, Edwards GE (1981) Photosynthetic capacity and nitrogen use efficiency of maize, wheat and rice: a comparison between C_3 and C_4 photosynthesis. J Experim Bot 32:459-466.

Schneider R (1982) Polychlorinated biphenyls (PCBs) in cod tissues from the western Baltic; significance of equilibrium partitioning and lipid composition in the bioaccumulation of lipophilic pollutants in gill-breathing animals. Meeresf Rep Mar Res (Ber D Wiss Kom Meeresf) 29:69-79.

Schulz-Bull DE, Petrick G, Kannan, Duinker JC (1995) Distribution of individual chlorobiphenyls (PCB) in solution and suspension in the Baltic Sea. Mar Chem 48:245-270.

Sengbusch P von (1989) Photosynthese. In: Botanik, McGraw Hill Book, Hamburg, Chapter 24:346-353.

Sericano JL, Wade TL, Jackson TJ, Brooks JM, Tripp BW, Farrington JW, Mee LD, Readman JW, Villeneuve JP, Goldberg ED (1995) Trace organic contamination in the Americas: an overview of the US national status & trends and the international 'Mussel Watch' programmes. Mar Poll Bull 31:214-225.

Shiskina OV, Pavlova GA, Bikova VS (1969) The geochemistry of halogens in the sea sediments and interstitial water (in Russian) Akad Nauk SSSR Inst Oceanol, Moscow, pp 117.

Sholkovitz ER, Cochran JK, Carey AE (1983) Laboratory studies of the diagenesis and mobility of ^{239,240}Pu and ^{137}Cs in nearshore sediments. Geochim Cosmochim Acta 47:1369-1379.

Sholkovitz ER, Mann DR (1984) The pore water chemistry of ^{239,240}Pu and ^{137}Cs in sediments of Buzzards Bay, Massachusetts. Geochim Cosmochim Acta 48:1107-1114.

Sijm DTHM, Seunen W, Opperhuizen A (1992) Life-cycle biomagnification study in fish. Environ Sci Technol 26: 2162-2174.

Sioud K (1994) Transfert de métaux entre eau et suspensions dans les estuaires. Thèse, L'Ecole Nationale des Ponts et Chaussées, Paris. pp 220.

Sivintsev Y (1994) Study of nuclides composition and characteristics of fuel in dumped submarine reactors and atomic icebreaker 'LENIN' Part 1 and Part 2. In: Working Material of the International Arctic Seas Assessment Project. IAEA, Vienna.

Sjoeblom K-L, Linsley GS (1995) Marine scientists on the Arctic Seas: documenting the radiological record. IAEA Bulletin, Vienna Austria, 37:25-30.

Skirrow G (1965) The dissolved gases - carbon dioxide, In: Riley JP, Skirrow G (eds) Chemical Oceanography, 1rst ed., Acad. Press, London, vol I:227-322.

Soulez-Larivière C, Le Péchon JC (1991) ESA standardisation process through the example of manned spacecraft atmosphere. In: Space environmental and control systems. Proc 4th Europ Symp, 21-24 Oct, ESA SP-324:129-131.

Stumm W, Morgan JJ (1970) Aquatic Chemistry, Wiley-Interscience, New York, pp 583.

Strand P, Nikitin A, Rudjord AL, Salbu B, Christensen G, Foyn L, Kryshev II, Chumichev VB, Dahlgaard H, Holm E (1994) Survey of artificial radionuclides in the Barents Sea and the Kara Sea. J of Environ Radioactivity 25:99-112.

Stumm W, Huang CP, Jenkins SR (1970) Specific chemical interaction affecting the stability of dispersed system Croat Chem Acta 42:223-245.

Stumm W, Morgan JJ (1981) Aquatic chemistry. Wiley-Intersc New York, pp 780.

Sundquist ET (1993) The global carbon dioxide budget. Science 259:934-941.

Sverdrup HU, Johnson MW, Flemin RH (1970) The oceans, their physics, chemistry and general biology (2nd ed). Prentice-Hall Inc, Englewood Cliffs, NJ, pp 1085.

Swift DJ (1989) The accumulation and retention of ^{95m}Tc by the edible winkle (*Littorina lirrorea* L.). J Environ Radioact 9: 31-52.

Tanabe S (1988) PCB problems in the future: foresight from current knowledge. Environ Pollut 50:5-28.

Tankere, SPC, Morley, NH, Burton, JD (1995) Spatial and temporal variations in concentrations of trace metals in the regions of the Straits of Sicily and Gibraltar. In: Martin J-M, Barth H (eds). Water Poll Res Rep 32 EROS 2000, Rep. EUR 16130 EN, 205-219.

Taube M (1992) Evolution of matter and energy. In: Hutzinger O (ed) The Natural Environment and the Biogeochemical Cycles, The Handbook of Environmental Chemistry, vol. 1, part F, Springer, Heidelberg, Berlin, New York, 65-182.

Thein M, Ballestra S, Yamato A, Fukai R (1980) Delivery of transuranic elements by rain to the Mediterranean Sea. Geochim Cosmochim Acta 44:1091-1097.

Tolosa I, Readman JW, Fowler SW, Villeneuve JP, Dachs J, Bayona JM, Albaiges J (1996) PCBs in the Western Mediterranean. Temporal trends and mass balance assessment. Deep Sea Res (in press).

Ullman WJ, Aller RC (1982) Diffusion coefficients in nearshore marine sediments. Limnol Oceanogr 27:552-556.

Villeneuve JP (1986) Géochemie des composés organochlorés dans l'environnement marin. Ph.D. Thesis Univ Paris VI. pp 180.

Volchok HL, Bowen VT, Folsom TR, Broecker WS, Schuert EA, Bien GS (1971) Oceanic distributions of radionuclides from nuclear explosions. In: Radioactivity in Marine Environment, US Nat. Acad. Sci.N.R.C., Washington, DC, 42-89.

Vrie EM van de, Duursma EK (1986) Residence times of contaminants in a lake-canal system: A delta case study in the SW Netherlands. Mar Chem 18:171-188.

Watson AJ, Law CS, Van Scoy KA, Millero FJ, Yao W, Friederich GE, Liddicoat MI, Wanninkhof RH, Barber RT, Coale KH (1994) Minimal effect of iron fertilization on sea-surface carbon dioxide concentrations. Nature 371:143-145.

Weiss RF (1974) Carbon dioxide in water and seawater: the solubility of a non-ideal gas. Mar Chem 2:203-215.

Werger MJA, Ellis RP (1981) Photosynthetic pathways in the arid regions of South Africa. Flora 171:64-75.

White Book 3 (1993) Facts and problems related to radioactive waste disposal in seas adjacent to the territory of the Russian Federation. Office of the President of the Russian Federation, Moscow.

Whitfield M, Turner DR (1979) Water-rock partition coefficients and the composition of seawater and river water. Nature 278:132-137.

Wollast R, Vanderborght JP (1994) Aquatic carbonate systems: chemical processes in natural waters and global cycles. In: Bidoglio G, Stumm W (eds) Chemistry of Aquatic systems: local and global perspectives, Kluwer Acad. Publ., Dordrecht pp 22.

World Bank (1993) Environmental screening. Environmental Assessment Sourcebook Update No. 2. Environmental Department, World Bank Washington DC USA.

Yefimov E (1994) Radionuclides composition, characteristics of shielding barriers and analyses of weak points of the dumped reactors of Submarine N 601. In: Working Material of the International Arctic Seas Assessment Project. IAEA, Vienna.

Yoon YY, Martin JM, Cotte MH (1995) Dissolved trace metals in the Western Mediterranean Sea: total concentration and speciation. In: Martin J-M, Barth H (eds). Water Poll Research Rep 32 EROS 2000, Rep. EUR 16130 EN, 235-250.

Zimmerman JTF (1976) Mixing and flushing of tidal embayments in the Western Dutch Wadden Sea Part I. Distribution of salinity and calculation of mixing time scales. Neth J Sea Res 10:149-191. Part II. Analysis of mixing processes. Neth J Sea Res 10:397-439.

Subject Index

Springer-Verlag
and the Environment

We at Springer-Verlag firmly believe that an international science publisher has a special obligation to the environment, and our corporate policies consistently reflect this conviction.

We also expect our business partners – paper mills, printers, packaging manufacturers, etc. – to commit themselves to using environmentally friendly materials and production processes.

The paper in this book is made from low- or no-chlorine pulp and is acid free, in conformance with international standards for paper permanency.

Printing: Mercedesdruck, Berlin
Binding: Buchbinderei Lüderitz & Bauer, Berlin